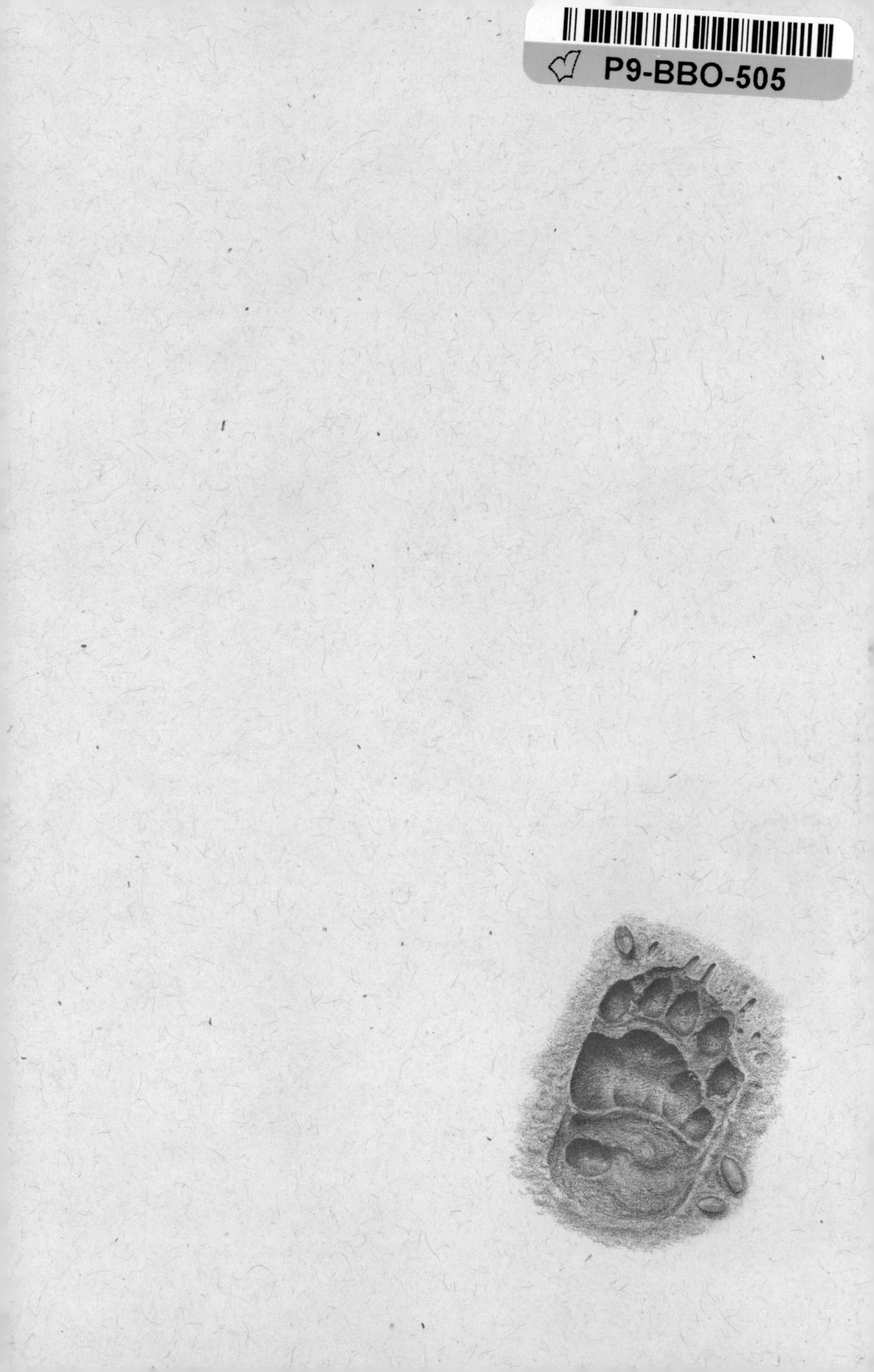

Small Wonder

BY THE SAME AUTHOR

FICTION

The Bean Trees

Pigs in Heaven

Animal Dreams

Homeland and Other Stories

The Poisonwood Bible

Prodigal Summer

NONFICTION

High Tide in Tucson: Essays from Now or Never

Holding the Line:
Women in the Great Arizona Mine Strike of 1983

POETRY

Another America

Barbara Kingsolver

Small Wonder

Illustrations by Paul Mirocha

HarperCollins*Publishers*

HarperCollins books may be purchased for educational, business, or sales promotional use. For information, please write: Special Markets Department, HarperCollins Publishers Inc., 10 East 53rd Street, New York, NY 10022.

An extension of this copyright page appears on page 269.

FIRST EDITION

Designed by Elliott Beard

Printed on acid-free paper

Library of Congress Cataloging-in-Publication Data is available upon request.

ISBN 0-06-050407-2

02 03 04 05 06 ❖/RRD 10 9 8 7 6 5 4 3 2 1

To treat life as less than a miracle is to give up on it.

WENDELL BERRY

Contents

Illustrated Catalog of Wonders

Foreword

I learned a surprising thing in writing this book. It is possible to move away from a vast, unbearable pain by delving into it deeper and deeper—by "diving into the wreck," to borrow the perfect words from Adrienne Rich. You can look at all the parts of a terrible thing until you see that they're assemblies of smaller parts, all of which you can name, and some of which you can heal or alter, and finally the terror that seemed unbearable becomes manageable. I suppose what I am describing is the process of grief.

I began this book, without exactly knowing I was doing so, on September 12, 2001. Someone from a newspaper had asked me to write a response to the terrorist attacks on the United States the day before. When you ask a novelist for a response, especially to something so immensely horrible, you had better sit down and wait awhile for the finish. I wrote my piece, then another one, and another. Sometimes writing seemed to be all that kept me from falling apart in the face of so much death and anguish, the one alternative to weeping without cease. Within a month I had published five different responses to different facets of a huge event in our nation's psychology—little pieces that helped me see the thing whole and try to bear it. I kept going. Soon I understood that I was examining aspects of life that seemed a world away from the World Trade Center towers or the Pentagon, but a world away is exactly where this grief begins and ends. This is a collection of essays about who we seem to be, what remains for us to live for, and what I believe we could make of ourselves. It began in a moment but ended with all of time.

The years since I last published an essay collection, in 1995, have been important ones in the ways of the world, and also in the ways of my family. I've given birth to my second child; the statistical makeup of the earth's population has moved from being mostly rural to mostly urban; wars have ended and wars have begun. I've been moved to write about each of these things, with increasing urgency, while taking into account all the others. Most of these essays are very new, but some have been published before in a different form. Three of them—"The Patience of a Saint," "Seeing Scarlet," and "Called Out"—were originally cowritten with my husband, Steven Hopp, as assignments for natural history magazines. These and other essays that began as short op-ed or magazine pieces inhaled and expanded to new girths when I offered them the chance to appear in a book. All of the previously published pieces have been partly or largely rewritten to take into account more recent events and to allow them to fit properly side by side in a collection. Some anomalies remain from their disparate origins; notably, I refer often to my children as my toddler, my kindergartner, my ten-year-old, and so on, as if I actually had an infinite number of offspring spanning all ages from birth to about fifteen. Strictly speaking, I have only two children, and this book is not about them; they just happened to be standing nearby while I looked for illumination, and so they cast their moving shadows.

This book isn't meant to be a commentary on specific political policies, though inevitably the day's headlines, like my children and all the other notable stuff of my life, have provided for me anecdotal entry into issues of more general and enduring interest. The several pieces that open and close the book respond most directly to current events, while many of the others form a collection of parables and reveries on parts of the world that may seem at first very distant from the epicenters of global crisis—a village at the edge of a Mexican jungle, for example, or my daughter's chicken coop. I ask the reader to understand that these essays are not incidental. I believe our largest problems have grown from the earth's remotest corners as well as our own backyards, and that salvation may lie in those places, too.

Compiling this book quickly in the strange, awful time that dawned on us last September became for me a way of surviving that time, and in the process I reopened in my own veins the intimate connection between the will to survive and the need to feel useful to something or someone beyond myself. In fact, that is a theme that runs through the book. Writing, which was both painful and palliative for me, turned out to be my own way of giving blood in a crisis. I can only hope this unit of words will have a longer shelf life than the forty-two days of a unit of blood, as this critical time blends seamlessly into the next one. I have tried to address, ultimately, things that don't rapidly change.

Some of the books that helped inform my writing here would constitute a good reading list for the new millennium: *Guns, Germs and Steel,* by Jared Diamond; *Eco-Economy: Building an Economy for the Earth,* by Lester R. Brown; *Earth in the Balance,* by Al Gore; *Shattering: Food, Politics and the Loss of Genetic Diversity,* by Cary Fowler and Pat Mooney; *Stolen Harvest: The Hijacking of the Global Food Supply,* by Vandana Shiva; *No Logo,* by Naomi Klein; *When Corporations Rule the World,* by David C. Korten; *Open Veins of Latin America,* by Eduardo Galeano; *Blowback, The Costs and Consequences of American Empire,* by Chalmers Johnson; *This Organic Life,* by Joan Dye Gussow; and anything by Wendell Berry.

Royalties from this book will help support the work of Physicians for Social Responsibility, Habitat for Humanity, Environmental Defense, and the humanitarian-aid project called Heifer International. I thank you on their behalf for the donation you've made, and encourage your continued support. Internet addresses for these organizations appear in the acknowledgments.

I dedicate this book to every citizen of my country who has suffered bereavement with honor, trepidation without panic, and the insult of fundamentalist condemnation without succumbing to similar thinking in turn. We may yet show the world we are worth our salt.

Small Wonder

Small Wonder

On a cool October day in the oak-forested hills of Lorena Province in Iran, a lost child was saved in an inconceivable way. The news of it came to me as a parable that I keep turning over in my mind, a message from some gentler universe than this one. I carry it like a treasure map while I look for the place where I'll understand its meaning.

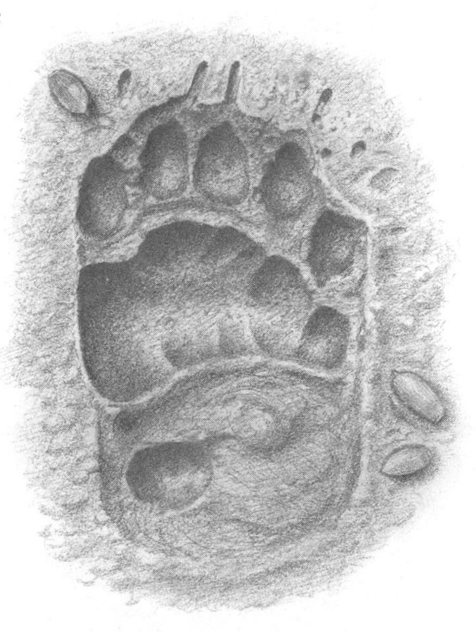

I picture it happening this way: The story begins with a wife and husband, nomads of the Lori tribe near Kayhan, walking home from a morning's work in their wheat. I imagine them content, moving slowly, the husband teasing his wife as she pulls her shawl across her face, laughing, and then suddenly they're stopped cold by the sight of a slender figure hurrying toward them: the teenage girl who was left in charge of the babies. In tears, holding her gray shawl tightly around her, she runs to meet the parents coming home on the road, to tell them in frightened pieces of sentences that he's disappeared, she has already looked everywhere, but he's gone. This girl is the neighbor's daughter, who keeps an eye on all the little ones too small to walk to the field, but now she has to admit wretchedly that their boy had strong enough legs to wander off while her attention was turned to—what? Another crying child, a fascinating insect—a thousand things can turn the mind from this to that, and the world is lost in a heartbeat.

They refuse to believe her at first—no parent is ever ready for this—and with fully expectant hearts they open the door flap of their yurt and peer inside, scanning the dim red darkness of the rugs on the walls, the empty floor. They look in his usual hiding places, under a pillow, behind the box where the bowls are kept, every time expecting this game to end with a laugh. But no, he's gone. I can feel how their hearts slowly change as the sediments of this impossible loss precipitate out of ordinary air and turn their insides to stone. And then suddenly moving to the fluttering panic of trapped birds, they become sure there is still some way out of this cage—here my own heart takes up that tremble as I sit imagining the story. Once my own child disappeared for only minutes that grew into half an hour, then an hour, and my panic took such full possession of my will that I could not properly spell my name for the police. But I could tell them the exact details of my daughter's eyes, her hair, the clothes she was wearing, and what was in her pockets. I lost myself utterly while my mind scattered out, carrying nothing but the search image that would locate and seize my child.

And that is how two parents searched in Lorena Province.

First their own village, turning every box upside down, turning the neighbors out in a party of panic and reassurances, but as they begin to scatter over the rocky outskirts it grows dark, then cold, then hopeless. He is nowhere. He is somewhere unsurvivable. A bear, someone says, and everyone else says No, *not* a bear, don't even say that, are you mad? His mother might hear you. And some people sleep that night, but not the mother and father, the smallest boys, or the neighbor's daughter who lost him, and early before the next light they are out again. Someone is sent to the next village, and larger parties are organized to comb the stony hills. They venture closer to the caves and oak woods of the mountainside.

Another nightfall, another day, and some begin to give up. But not the father or mother, because there is nowhere to go but this, we all have done this, we bang and bang on the door of hope, and don't anyone dare suggest there's nobody home. The mother weeps, and the father's mouth becomes a thin line as he finds several men willing to go all the way up into the mountains. Into the caves. Five kilometers away. In the name of heaven, the baby is only sixteen months old, the mother tells them. He took his first steps in June, a few weeks before Midsummer Day. He can't have walked that far, everybody knows this, but still they go. Their feet scrape the rocky soil; nobody speaks. Then the path comes softer under the live oaks. The corky bark of the trees seems kinder than the stones. An omen. These branches seem to hold promise. Lori people used to make bread from the acorns of these oaks, their animals feed on the acorns, these trees sustain every life in these mountains—the wild pigs, the bears. Still, nobody speaks.

At the mouth of the next cave they enter—the fourth or the hundredth, nobody will know this detail because forever after it will be the first and last—they hear a voice. Definitely it's a cry, a child. Cautiously they look into the darkness, and ominously, they smell bear. But the boy is in there, crying, alive. They move into the half-light inside the cave, stand still and wait while the smell gets danker and the texture of the stone walls weaves its details more clearly into their vision. Then they see the animal, not a dark hollow in the cave wall as they first thought but the dark, round shape of a thick-furred, quiescent she-bear lying against the wall. And then they see the child. The bear is curled around him, protecting him from these fierce-smelling intruders in her cave.

I don't know what happened next. I hope they didn't kill the bear but instead simply reached for the child, quietly took him up, praised Allah and this strange mother who had worked His will, and swiftly left the cave. I've searched for that part of the story—whether they killed the bear. I've gone back through news sources from river to tributary to rivulet until I can go no further because I don't read Arabic. This is not a mistake or a hoax; this happened. The baby was found with the bear in her den. He was alive, unscarred, and perfectly well after three days—and well fed, smelling of milk. The bear was nursing the child.

What does it mean? How is it possible that a huge, hungry bear would take a pitifully small, delicate human child to her breast rather than rip him into food? But she was a mammal, a mother. She was lactating, so she must have had young of her own somewhere—possibly killed, or dead of disease, so that she was driven by the pure chemistry of maternity to take this small, warm neonate to her belly and hold him there, gently. You could read this story and declare "impossible," even though many witnesses have sworn it's true. Or you could read this story and think of how warm lives are drawn to one another in cold places, think of the unconquerable force of a mother's love, the fact of the DNA code that we share in its great majority with other mammals—you could think of all that and say, Of course the bear nursed the baby. He was crying from hunger, she had milk. Small wonder.

The story of the child and the bear came to me on the same day I read the year's opening words on the bombing campaign in Afghanistan. I sat very still at the table that morning while my coffee went cold and my eyes scanned one sentence after another, trying to absorb the account of explosives raining from the sky on a place already ruled by terror, by all accounts as poor and war-scarred a populace as has ever crept to a doorway and looked out. My heart was already burdened by grief; only days had passed since I sat in this same place, at the same time of day, and listened to a report that unfolded unbelievably, numbingly, into a litany of unimaginable terror and assault on this country that holds my love and life. I could hardly bear more. But now my mind's eye ran away to find women on the other side of the world who were looking just then from their children up to the harrowing skies. What would they make of this message, whose retaliatory import seemed so perfectly clear to us? I read that the bombs had taken, among others, four humanitarian-aid workers in the small office in Kabul that coordinated the work of removing land mines from the soil of that beleaguered nation. My heart's edge felt as dull and pocked as an old shovel as it scooped low to take on this new weight, the rubble and grief of war. And so when I came to the opposite page in the book of miracles, I cleaved hard to this other story. People not altogether far away from Kabul—wrapping themselves in similar soft robes, similar hopes—had been visited by an impossible act of grace.

In a world whose wells of kindness seem everywhere to be running dry, a bear nursed a lost child. The miracle of Lorena is genuine. If you venture onto the information highway with a good search engine and propose "Kayhan, Iran, bear," you will find this tiny, remarkable note in the human archive. You may also find, as I did, a report written by Heshmatollah Tabarzadi, telling how he and eleven other Iranian students heading home after a rally were arrested and tortured for protesting against government oppression. His story comes up in the search because he also lived near Kayhan, and buried deep in the text are the words of an officer who told him during one of his strenuous torture sessions, "We can milk roosters here, bears lay eggs here, you?! You're just a human being. In the course of one hour we can make a bear confess to being a rabbit." Another small footnote in the human archive. God is frightful, God is great—you pick. I choose this: God is in the details, the completely unnecessary miracles sometimes tossed up as stars to guide us. They are the promise of good fortune in a cloudless day, and the animals in clouds; look hard enough, and you'll see them. Don't ask if they're real.

I elect to believe that the Lori men didn't kill the bear. For years to come I will picture the father quietly lifting the boy from her belly, wrapping him in the soft cloth of his shirt, and reverently leaving the cave of his salvation. Leaving a small pile of acorns outside the lair of this mother, this instrument of Allah's design, as a sacrament.

I believe in parables. I navigate life using stories where I find them, and I hold tight to the ones that tell me new kinds of truth. This story of a bear who nursed a child is one to believe in. I believe that the things we dread most can sometimes save us. I am losing faith in such a simple thing as despising an enemy with unequivocal righteousness. A mirror held up to every moral superiority will show its precise mirror image: The terrorist loves his truth as hard as I love mine; he has a mother who looks on her child with the same fierce pride I feel when I look at my own. Someone, somewhere, must wonder how I could love the boys who dropped the bombs that killed the humanitarian-aid workers in Kabul. We are all beasts in this kingdom, we have killed and been killed, and some new time has come to us in which we are called out to find another way to divide the world. Good and evil cannot be all there is.

Lately we've had to consider a new kind of enemy we can

hardly bear to behold: a foul hatred bent to the destruction of all things precious to us—to *me;* I'll shudder here and speak for myself as someone who loves her life as it is, a woman whose spirit would surely get itself stoned to death if forced to submit to the order of such men. The horrors they've wrought have reduced me at times to a pure grief in which I could only cross my arms against my chest and cry out loud. I can't pretend to understand their aims; I can barely grasp the motives of a person who hits a child, so I surely have no access to the minds of men who could slaughter thousands of innocents and die in the process, or train others to do those things. I presume they want us to become more like themselves: hateful, self-righteous, violent. I expect they would count themselves victorious to see us reduced to panic under their specter, to fall into factions of difference and censor or attack our own minorities, to weaken and let go of the ideals of equality and kindness that first brought our country onto the map of the world. So I hold my own heart fast against the fulfillment of this horrific prophecy, and I hope that the men who constructed it can be made to live out humiliated ends in prison—a punishment that would inspire fewer followers, I think, than dramatic death in battle.

But even that would not be the end of the story. This new enemy is not a person or a place, it isn't a country; it is a pure and fearsome ire as widespread as some raw element like fire. I can't sensibly declare war on fire, or reasonably pretend that it lives in a secret hideout like some comic-book villain, irrationally waiting while my superhero locates it and then drags it out to the thrill of my applause. We try desperately to personify our enemy in this way, and who can blame us? It's all we know how to do. Declaring war on a fragile human body and then driving the breath from it—that is how enmity has been dispatched for all of time, since God was a child and man was even more of one.

But now we are faced with something new: an enemy we can't

kill, because it's a widespread anger so much stronger than physical want that its foot soldiers gladly surrender their lives in its service. We who live in this moment are not its cause—instead, a thousand historic hungers blended to create it—but we are its chosen target: We threaten this hatred, and it grows. We smash the human vessels that contain it, and it doubles in volume like a magical liquid poison and pours itself into many more waiting vessels. We kill its leaders, and they swell to the size of martyrs and heroes, inspiring more martyrs and heroes. This terror now requires of us something that most of us haven't considered: how to defuse a lethal enemy through some tactic more effective than simply going at it with the biggest stick at hand.

Something new is upon us, and yet nothing is ever new. Two thousand years ago, the Greeks understood this enemy. It inhabited their imagination in a favorite tale of heroic adversity, the story of Jason and the Argonauts, where it took the shape of a particular dragon. When this creature was slain and its corpse fell to the soil, each of its teeth germinated and instantly grew into a new enemy, fully armed and born full-force to the battle. In all of his picturesque predicaments, Jason never faced one more impossible than this field of foes, each with its own mouthful of teeth aching to germinate. For once he couldn't fight his way out; it took a woman to save him. Medea, who loved Jason and tried to protect him, whispered in his ear a simple truth: Hatred dies only when turned on itself. This force could not be extinguished by the sword, she told him; only a clever psychological strategy could vanquish it. Jason took her advice but went about it in his own way, by throwing a rock cryptically and inciting an internal riot of rock throwing in which the dragon's-tooth warriors destroyed one another.

Later on, Jason encountered another dragon. Unbelievably (but of course, heroes being only what they are, predictably), he again drew his sword, ready to kill it. Medea stopped him with a quick, gentle hand on his bicep. This time, rather than allowing a new field of hatred to be sown and reaped, she moved quietly to the mouth of the sleeping dragon and gave it an elixir of contentment so it would remain asleep as she and her lover passed by.

At a time when the modern imagination seems fully engaged in discussion of swords of every length and breadth, there's little room for other kinds of talk. But I'm emboldened by Medea to speak up on behalf of psychological strategy. It's not a simple-minded suggestion; her elixir of contentment is exactly as symbolic as Jason's all-conquering sword, and the latter has by no means translated well into reality. The strategic difference is the capacity to understand this one thing: Some forms of enemy are made more deadly by killing. It would require the deepest possible shift of our hearts to live in this world of fundamental animosity and devote ourselves not to the escalating exertion to kill, but rather, to lulling animosity to sleep. Modern humanity may not be up to the challenge. Modern humanity may not have a choice.

The miracle of Lorena Province haunts me as I consider this predicament. I catch glimpses of that bear pacing restlessly on the periphery of everything I thought I could be sure of. We are alive in a fearsome time, and we have been given new things to fear. We've been delivered huge blows but also huge opportunities to reinforce or reinvent our will, depending on where we look for honor and how we name our enemies. The easiest thing is to think of returning the blows. But there are other things we must think about as well, other dangers we face. A careless way of sauntering across the earth and breaking open its treasures, a terrible dependency on sucking out the world's best juices for ourselves—these may also be our enemies. The changes we dread most may contain our salvation. And the stinging truth that we aren't entirely loved for our ways in this world? Like the bear, this thing could eat us up or save us. We will see.

There are many angles on the miracle in Lorena: for one, that a bear was in the cave at all. Bears are scarce in the world now, relative to their numbers in times of old; they're a rare sight even in the wildest mountains of Iran. They have been hunted out and nearly erased from the mountains and forests of Europe, much of North America, and other places that have been inhabited for thousands of years by humans, who by and large find it difficult to leave large predators alive. Bears and wolves are our fairy-tale archenemies, and in these tales we teach our children only, and always, to kill them, rather than to tiptoe past and let them sleep. Maybe that's why I'm comforted by the image of a small child curled in the embrace of a mother bear. We need new bear and wolf tales for our times, since so many of our old ones seem to be doing us no good. Now we're finding that it takes our every effort of will and imagination to pull back, to stop in our tracks as hunter and hunted, to halt our habit of killing, before every kind of life we know arrives at the brink of extinction.

Some days you have to work hard to save the bear. Some days the bear will save you.

> *Something there is that doesn't love a wall,*
> *That sends the frozen-ground-swell under it*
> *And spills the upper boulders in the sun. . . .*

In his poem "Mending Wall," Robert Frost invokes the image of his neighbor walking the fence line intent on constant survey and repair, here and there raising up a boulder between his hands, "like an old-stone savage armed," to put it back in its place, determined to keep this boundary intact, though it restrains only trees. (*My apple trees will never get across / And eat the cones under his pines, I tell him.*) "Good fences make good neighbors," is the only rationale the neighbor will offer, as his father said before him. The poet is baffled at so much resolute effort.

And so we all might well feel baffled, as we awaken this morning to find the greatest part of our ways and means invested in the walls our nations have built between ourselves and those whom we wish to keep out. Throughout our modern history we have taken each step in the construction of defensive borders with few doubts in mind, from stones to bricks and mortar, to rifles and barbed wire, to missiles and tanks and the firestorm contained in an atom. And now here we are, devoted to the efforts of surveillance, repair, and dread.

Borders crumble; they won't hold together on their own; we have to shore them up constantly. They are fortified and patrolled by armed guards, these fences that divide a party of elegant diners on one side from the children on the other whose thin legs curve like wishbones, whose large eyes peer through the barbed wire at so much food—there is no wall high enough to make good in such a neighborhood. For this, of course, is what the fences divide. Probably we began with more theoretical notions of ethnic purity—the wish to keep the apples out of our pines—and for most of the last century we rationalized our walls in terms of ideology, but the Iron Curtain has now dramatically fallen. Now we have fashioned from the crumbling boundaries of the Cold War a whole new shape of division, fundamentally between rich and poor. That chasm keeps growing; a quarter of the world's poor are now poorer than they were fifteen years ago, having struggled only to lose ground.

The hard boundary between the haves and the have-nots is still defended with armaments, but now it is also bridged by a dancing, illusory world of material wants. Passing through every wall are electronic beams that create a shadow play of desire staged by the puppeteers of globalized commerce, who fund their advertising each year with more than a hundred dollars spent for this planet's every man, woman, child. "This world of inequality is also a world of solitude," writes Eduardo Galeano, in which multitudes of the desperate are led "to confuse being with having." And condemned to have not.

In the name of God and all the fishes, a hundred dollars for every human alive, solely to lure them all into want! To consider this material tyranny is to begin, surely, to understand why protest against the global corporate order throws itself down weeping in the streets from Seattle to Genoa to Pakistan. Imagine how it looks from the other side, where undulating female bodies sell soft drinks to the likes of the nomads in Lorena Province. This omnipresent shadow play ignores the genuine differences of culture that legitimately distinguish us, pretending instead that we are all of one mind and share in an equal and endless pocketbook. But in truth, it relies on other kinds of fences, ever-increasing fortifications of the heart and treasury. Global commerce is driven by a single conviction: the inalienable right to earn profit, regardless of any human cost. No one argues that this is untrue. It is also proving unassailable. When Guatemala passed laws intended to encourage low-income mothers to breast-feed rather than use imported infant foods, for instance, the world's largest baby-food manufacturer circumvented these rules by threatening to invoke GATT and other trade sanctions. When Europeans favored importing bananas from the Caribbean and Canary Islands over buying fruit grown where working conditions are worse, they were similarly thwarted. The laws governing international trade render it more difficult each year to inject moral considerations into the marketplace, frustrating the many nations and individuals who still wish to balance economic motives with compassionate ones. Indeed, international trade laws increasingly restrict access to the very information that makes any such concession possible—witness, for example, the endless battle for accurate labeling waged by U.S. consumers who prefer their food organically grown and not genetically modified. The profiteering drive of commerce owns no malice or mercy, is incapable of regret, and takes no prisoners; it is simply an engine with no objective but to feed itself. And it is a Goliath: A decade ago, the combined sales of the world's ten largest corporations exceeded the gross national product of the world's hundred smallest countries *put together,* and the gap is growing.

Inevitably, hungry souls and angry hands rise up against that amoral giant, and ever-higher walls of armaments are required to keep them at bay. These walls create among us a huge class that the French author Jacques Attali has named the "millennial losers," for whom the fantasy of prosperity promoted by the media is both a continuous allure and an endless slapdown. The siren's song calls them toward Paris and New York, glittering Emerald Cities walled off by inaccessibility. In his 1991 book *Millennium: Winners and Losers in the Coming World Order,* Attali observed with a chilling prescience that particularly among those in the Middle East who'd suffered repeated humiliations by the West, the fiercely absent presence of worldly affluence tended to inspire fervent cults of frustration and outrage.

Something there is that doesn't love a wall,
That wants it down. . . .

We who are alive in this moment didn't build these walls, nor did we ignite the fury that has smoldered for eons and hurls itself at us now as a burning question. But we have inherited the urgent necessity of answering it. And possibly we will succeed.

It isn't only the insane rage of the dispossessed that burns against these walls. It is also, from inside the house of privilege, the indignation of the children of mercy who laid themselves down in the streets of Seattle to bring the World Trade Organization's autocracy to a halt. It is the shaken hearts of those who gathered at

Ground Zero in Manhattan and begged for an end to the killing fields. It is surely every person's animal soul, and the DNA code shared perfectly between every two humans alive—a genetic truth inside our cells that leaps back and forth across continents. That code insists on my kinship with both the elegant diner and the child with the wishbone legs, with the Lori nomad and even, perhaps, with the bear. Roosters give milk here, bears lay eggs. The lion could lie down with the lamb. A frozen groundswell just beyond our senses heaves and buckles, daring the world to dismantle these walls of enmity and use the stones to build ovens for baking bread. It would be the death of something, and the life of something. Somewhere there must be a door through. The alternative is only to construct higher walls, and the higher they grow, the harder they will fall. It's hard to imagine a more frightening time than this.

I know, someone has said that already. People said it a thousand years ago, and they've said it about nearly every minute that has ever gone by since. My generation's parents said it during the Cuban missile crisis, and *their* parents said it after Pearl Harbor. Mothers said it as they watched their sons ride off to fight in the Civil War, and they said it a hundred years later when black-skinned children had to be escorted by armed officers through the doors of an all-white school. The day Martin Luther King Jr. was murdered, or Gandhi, or Jesus, or Monseñor Oscar Romero, or the day the Buddhist monks immolated themselves in Vietnam while the stunned world watched—all of these were the worst there could ever be. Scholars of history are fond of pulling up statements of dismay from time immemorial to prove to us that there is nothing new under the sun: The wolf was always at the door, and people have always been hell-bent on pulverizing one another, exactly as they are now.

The historians are right, it isn't new, this feeling of despair over a world gone mad with heartless and punitive desires. It isn't new that both sides rush to the fundamentalist presumption of themselves against the evil ones. It isn't even new that the world could fall apart and become permanently uninhabitable in a matter of minutes—that's what the Cuban missile crisis was about. What is new is that we now know so very much about the world, or at least the part of it that is most picturesquely exploding on any given day, that we're left with a desperate sense that *all* of it is exploding, *all* the time. As far as I can tell, that is the intent and purpose of television news. We see so much, understand so little, and are simultaneously told so much about What We Think, as a populace polled minute by minute, that it begins to feel like an extraneous effort to listen at all to our hearts.

I try with all my might to duck under this wire, not to believe in polls or allow the TV bluster anywhere near my face. At moments I have to stop taking in more news so I can consider what I've gathered so far and pay attention to my own community, since that is the only place where I can muster a posse to take on our own local disasters of the day. Sometimes I have to make a simple, straightforward effort to do just that, so I will feel less like a screen door banging in a hurricane.

And that is what is really new since time immemorial: the sense that the problems are so vast in scope that we've lost any hope of altering the course of things. During previous eras of conspicuous doom—the Black Death, for example—people surely felt that the world was ending, but the end they probably pictured was smaller in scale, consisting of themselves, their neighbors, and God. They couldn't imagine a wreckage so appalling as the end of humankind on a planet made squalid by man's own hand; I doubt they had yet grasped the magnificence of our history, or the infinitude of our foolishness.

The feeling I dread most is not fear but despair—the dim, oppressive sense that the more things change, the more they stay the same; that each of us with a frozen heart "like an old-stone savage armed" will continue to move in darkness, lifting boulders, patrolling the firmaments of divisive anger. I do not go gentle into that particular night; I burn and rave against the dying of all hope. I concede that there is mounting evil in this world, and that some hearts are so hardened already that they cannot possibly be appeased. Some walls grow higher each year, it's true.

But others crumble. The people who said the sky would fall and God would weep if their sons and daughters had to sit in the same schoolroom as black-skinned children were wrong: The sky didn't, and whether or not God did is a matter of personal opinion. The earth has shifted beneath our feet, time and again, as the stones of our paradigms fell hard on the dust. Irrevocably, humanity inhaled a new era in 1772, when Lord Mansfield declared that slaves were free the moment they breathed the air of England; sixty years later that promise was extended to the air of the whole British Empire, and thirty-odd years after *that,* following a monstrous sacrifice of earnest belief on both sides, to the air of the United States. In the *next* century, at the end of which the women of Afghanistan fell from full citizenship to a nation of silent, peering eyes, there were many other countries—including mine—where women fasted and marched and fought for and gained the right to own property, then to vote, then to sit on a jury of their peers and be counted as fully human. Apartheid fell step by step in the United States and in South Africa. Mate choice and romantic love have come to be regarded, at least in some places, as private and sacrosanct privileges, limited only by the congenital rules of a complex human chemistry, even if the romance should cross lines of class or color or contradict the common presumptions of gender.

Some of these changes I've witnessed in my lifetime. I began first grade in a segregated public school, in a state whose anthem—which we children doggedly belted out each morning—contained the line "'Tis summer, the darkies are gay!" That unsettling declaration would be quietly revised many times, for many reasons, over the next ten years. Time and again, the bear they had sworn would rip us limb from limb was begrudgingly allowed a place at the table, and behold, it used a fork and a spoon. The natural laws we have believed in and taught our children have sometimes been found to be not natural laws at all, but rather fearsome constructs of our own making, undermined by the evidence. And among those mistakes there is this: All of the promises of politicians, generals, madmen, and crusaders that war can create peace have yet to be borne out.

With these startling honesties glinting up at us from history's broken mirror, it strikes me that this is worth shouting from the rooftops: We could be wrong this time, again. The enemy may not be exactly what we think. It may be a force that resides in many quarters, including inside our skin, in our very words, the questions we frame, the things we love most, the things we can't live without. Our greatest dread may be our salvation. We are in no position yet to declare the moral of our story.

"But how," a friend in New York has asked me, "do I live with the anger? On my street, where every night I smell the incinerated towers as I walk the dog, there is inescapable pain and rage. In this city we bear the brunt of the struggle to figure out what it means to have our hearts cut away by hateful violence that the whole country somehow engendered."

I hardly know how to answer that for myself, let alone for anyone else, and yet that struggle is for so many of us the currency of everyday survival. Most of us can't know how it feels to live in the shadow of those murdered towers. But nearly all of us care, because we ourselves know what it means to have our hearts cut away by life: We've borne physical assaults, lost those we loved best, lost our own souls, our physical wholeness, the future we were counting on, had days we could not see how to get through. I understand that we'll have lost everything if a hateful enemy can crush us and reconstruct us in its angry image, but what other door may lead out of this dark room? I've felt outrage I was sure would burn me alive. Some nights I've lain awake wondering how to keep on living while someone, somewhere, despises me and wishes so many of us dead because of our faith or nationality, assigning to us transgressions I can scarcely grasp. I wonder how to stay calm with so much beauty at stake, being scorched from my line of sight as trees fall and sacred places are ground to dust. I find it insufferable to bear silent witness to the flesh-and-bone devastations of war, and bitterly painful to be cast sometimes as a traitor to the homeland I love, simply because I raise questions. I find myself in a strange niche, reviled by some compatriots because I can't praise war as the best answer, and reviled everywhere else because my nation does. Each of us inhabits his or her own strange niche, I suppose; we've engendered animosity for many things that most of us never contrived to do, perhaps never even knew about. Many of us can't fully believe in all the imperatives that have been pronounced the will of our people. One problem with democracy as it plays in our country is that the majority rules so hard; we seem bent on dividing all things into a contest of Win and Lose, and declaring that the Losers are *losers.* Nearly half of us are routinely asked to disappear while the slim majority works its will. But the playing field is the planet earth, and I for one have no place else to go.

The closest my heart has come to breaking lately was on the day my little girl arrived home from school and ran to me, her face tense with expectation, asking, "Are they still having that war in Afghanistan?"

As if the world were such a place that in one afternoon, while kindergartners were working hard to master the letter *L,* it would decide to lay down its arms. I tried to keep the tears out of my eyes. I told her I was sorry, yes, they were still having the war.

She said, "If people are just going to keep doing that, I wish I'd never been born."

I sat on the floor and held her tightly to keep my own spirit from draining through the soles of my feet. I don't know what other mothers say at such moments; I suppose some promise that only the bad men are getting hurt. I wish I could believe in that story myself. But my children have never been people I could lie to. My best revenge against all the dishonesty and hatred in the world, it seems to me, will be to raise right up through the middle of it these honest and loving children.

I asked her, "Do you really mean that? You wish you'd never known Daddy or me or your sister? That you'd never gotten a chance to hug us, or have us read books to you, or tuck you in at night? Never gotten to take care of your chickens and gather their eggs, never seen a rainbow?"

Of course she said, soon enough, that she was glad to be alive. And I'm sure that's true, as I watch her throw her heart and limbs into a mostly unburdened life. But I understood that day that we are all in the same boat. It's the same struggle for each of us, and the same path out: the utterly simple, infinitely wise, ultimately defiant act of loving one thing and then another, loving our way back to life.

It used to be, on many days, that I could close my eyes and sense myself to be perfectly happy. I have wondered lately if that feeling will ever come back. It's a worthy thing to wonder, but maybe being perfectly happy is not really the point. Maybe that is only some modern American dream of the point, while the truer measure of humanity is the distance we must travel in our lives, time and again, "twixt two extremes of passion—joy and grief," as Shakespeare put it. However much I've lost, what remains to me is that I can still speak to name the things I love. And I can look for safety in giving myself away to the world's least losable things.

My parents, before raising me, first had to spend every day of their lives from infancy to early adulthood coping with a great depression and then a war. As a consequence, they reared me under the constant counsel to trust spiritual values ahead of material ones, and to look to the land for shelter. "A house can burn down," they said, "but a piece of land will always be there." These words came back to me profoundly on the bleakest day, when we watched those two shattered towers billow smoke.

I've internalized my parents' message in a way that is not precisely personal; after all, ownership of farms per se provided no safety for the Japanese Americans removed to concentration camps during World War II, many of whom lost everything. But I understand what they meant, and have spent a lifetime learning to believe in things that can never burn down. I can invest my heart's desire and the work of my hands in things that will outlive me. Although it grieves me that houses are burning, I have fallen in love with a river that runs through a desert, a rain forest at the edge of night, the right of a species to persist in its own wild place, and the words I might assemble to tell their stories. I've fallen in love with freedom regardless, and the entitlement of a woman to get a move on, equipped with boots that fit and opinions that might matter. The treasures I carry closest to my heart are things I can't own: the curve of a five-year-old's forehead in profile, and the vulnerable expectation in the hand that reaches for mine as we cross the street. The wake-up call of birds in a forest. The intensity of the light fifteen minutes before the end of day; the color wash of a sunset on mountains; the ripe sphere of that same sun hanging low in a dusty sky in a breathtaking photograph from Afghanistan.

In my darkest times I have to walk, sometimes alone, in some green place. Other people must share this ritual. For some I suppose it must be the path through a particular set of city streets, a comforting architecture; for me it's the need to stare at moving water until my mind comes to rest on nothing at all. Then I can go home. I can clear the brush from a neglected part of the garden, working slowly until it comes to me that here is one small place I can make right for my family. I can plant something as an act of faith in time itself, a vow that we will, sure enough, have a fall and a winter this year, to be followed again by spring. This is not an end in itself, but a beginning. I work until my mind can run a little further on its tether, tugging at this central pole of my sadness, forgetting it for a minute or two while pondering a school meeting next week, the watershed conservation project our neighborhood has undertaken, the farmer's market it organized last year: the good that becomes possible when a small group of thoughtful citizens commit themselves to it. And indeed, as Margaret Mead said, that is the only thing that ever really does add up to change. Small change, small wonders—these are the currency of my endurance and ultimately of my life. It's a workable economy.

Political urgencies come and go, but it's a fair enough vocation to strike one match after another against the dark isolation, when spectacular arrogance rules the day and tries to force hope into hiding. It seems to me that there is still so much to say that I had better raise up a yell across the fence. I have stories of things I believe in: a persistent river, a forest on the edge of night, the religion inside a seed, the startle of wingbeats when a spark of red life flies against all reason out of the darkness. One child, one bear. I'd like to speak of small wonders, and the possibility of taking heart.

Saying Grace

I never knew what *grand* really was until I saw the canyon. It's a perspective that pulls the busy human engine of desires to a quiet halt. Taking the long view across that vermilion abyss attenuates humanity to quieter internal rhythms, the spirit of ice ages, and we look, we gasp, and it seems there is a chance we might be small enough not to matter. That the things we want are not the end of the world. I have needed this view lately.

I've come to the Grand Canyon several times in my life, most lately without really understanding the necessity. As the holidays approached, I couldn't name the reason for my uneasiness. We thought about the cross-country trip we've usually taken to join our extended family's Thanksgiving celebration, but we didn't make the airplane reservations. Barely a month before, terrorist attacks had distorted commercial air travel to a horrifying new agenda, one that left everybody jittery. We understood, rationally, that it was as safe to fly as ever, and so it wasn't precisely nervousness that made us think twice about flying across the country for a long weekend. Rather, we were moved by a sense that this was wartime, and the prospect of such personal luxury felt somehow false.

I called my mother with our regrets and began making plans for a more modest family trip. On the days our daughters were out of school, we would wander north from Tucson to revisit some of the haunts I've come to love in my twenty years as a desert dweller transplanted from the verdant Southeast. We would kick through the leaves in Oak Creek Canyon, bask like lizards in the last late-autumn sun on Sedona's red rocks, puzzle out the secrets of the labyrinthine ruins at Wupatki, and finally stand on the rim of that remarkable canyon.

I felt a little sorry for myself at first, missing the reassuring tradition of sitting down to face a huge, upside-down bird and counting my blessings in the grand, joyful circle of my kin. And then I felt shame enough to ask myself, How greedy can one person be, to want more than the Grand Canyon? How much more could one earth offer me than to lay herself bare, presenting me with the whole of her bedrock history in one miraculous view? What feast could satisfy a mother more deeply than to walk along a creek through a particolored carpet of leaves, watching my children pick up the fine-toothed gifts of this scarlet maple, that yellow aspen, piecing together the picture puzzle of a biological homeplace? We could listen for several days to the songs of living birds instead of making short work of one big dead one. And we'd feel lighter afterward, too.

These are relevant questions to ask, in this moment when our country demands that we dedicate ourselves and our resources, again and again, to what we call the defense of our way of life: How greedy can one person be? How much do we need to feel blessed, sated, and permanently safe? What is safety in this world, and on what broad stones is that house built?

Imagine that you come from a large family in which one brother ended up with a whole lot more than the rest of you. Sometimes it happens that way, the luck falling to one guy who didn't do that much to deserve it. Imagine his gorgeous house on a huge tract of forests, rolling hills, and fertile fields. Your other relatives have decent places with smaller yards, but yours is mostly dust. Your lucky brother eats well, he has meat every day—in fact, let's face it, he's corpulent, and so are his kids. At your house, meanwhile, things are bad: Your kids cry themselves to sleep on empty stomachs. Your brother must not be able to hear them from the veranda where he dines, because he throws away all the food he can't finish. He will do you this favor: He's made a TV program of himself eating. If you want, you can watch it from your house. But you can't have his food, his house, or the car he drives around in to view his unspoiled forests and majestic purple mountains. The rest of the family has noticed that all his driving is kicking up dust, wrecking not only the edges of his property but also their less pristine backyards and even yours, which was dust to begin with. He's dammed the river to irrigate his fields, so that only a trickle reaches your place, and it's nasty. You're beginning to see that these problems are deep and deadly, that you'll be the first to starve, and the others will follow. The family takes a vote and agrees to do a handful of obvious things that will keep down the dust and clear the water—all except Fat Brother. He walks away from the table. He says God gave him good land and the right to be greedy.

The ancient Greeks adored tragic plays about families like this, and their special word for the Fat Brother act was *hubris*. In the town where I grew up we called it "getting all high and mighty," and the sentence that came next usually included the words "getting knocked down to size." For most of my life I've felt embarrassed by a facet of our national character that I would have to call prideful wastefulness. What other name can there be for our noisy, celebratory appetite for unnecessary things, and our vast carelessness regarding their manufacture and disposal? In the autumn of 2001 we faced the crisis of taking a very hard knock from the outside, and in its aftermath, as our nation grieved, every time I saw that wastefulness rear its head I felt even more ashamed. Some retailers rushed to convince us in ads printed across waving flags that it was our duty even in wartime, *especially* in wartime, to get out and buy those cars and shoes. We were asked not to think very much about the other side of the world, where, night after night, we were waging a costly war in a land whose people could not dream of owning cars or in some cases even shoes. For some, "wartime" became a matter of waving our pride above the waste, with slogans that didn't make sense to me: "Buy for your country" struck me as an exhortation to "erase from your mind what just happened." And the real meaning of this one I can't even guess at: "Our enemies hate us because we're free."

I'm sorry, but I have eyes with which to see, and friends in many places. In Canada, for instance, I know people who are wicked cold in winter but otherwise in every way as free as you and me. And nobody hates Canada.

Hubris isn't just about luck or wealth, it's about throwing away food while hungry people watch. Canadians were born lucky, too, in a global sense, but they seem more modest about it, and more deeply appreciative of their land; it's impossible to imagine Canada blighting its precious wilderness areas with "mock third-world villages" for bombing practice, as our air force has done in Arizona's Cabeza Prieta Range. I wonder how countries bereft of any wild lands at all view our plans for drilling in the Arctic National Wildlife Refuge, the world's last immense and untouched wilderness, as we stake out our right to its plunder as we deem necessary. We must surely appear to the world as exactly what we are: a nation that organizes its economy around consuming twice as much oil as it produces, and around the profligate wastefulness of the wars and campaigns required to defend such consumption. In recent years we have defined our national interest largely in terms of the oil fields and pipelines we need to procure fuel.

In our country, we seldom question our right to burn this fuel in heavy passenger vehicles and to lead all nations in the race to pollute our planet beyond habitability; some of us, in fact, become belligerent toward anyone who dares to raise the issue. We are disinclined as a nation to assign any moral value at all to our habits of consumption. But the circle of our family is large, larger than just one nation, and as we arrive at the end of our frontiers we can't possibly be surprised that the rest of the family would have us live within our means. Safety resides, I think, on the far side of endless hunger. Imagine how it would feel to fly a flag with a leaf on it, or a bird—something *living*. How remarkably generous we could have appeared to the world by being the first to limit fossil-fuel emissions by ratifying the Kyoto agreements, rather than walking away from the table, as we did last summer in Bonn, leaving 178 other signatory nations to do their best for the world without any help from the world's biggest contributor to global warming. I find it simply appalling that we could have done this; I know for a fact that many, many Americans were stunned, like me, by the selfishness of that act, and can hardly bear their own complicity in it. Given our societal devotion to taking in more energy than we put out, it's ironic that our culture is so cruelly intolerant of overweight individuals. As a nation we're not just overweight (a predicament that deserves sympathy); I fear we are also, as we live and breathe, possessed of the Fat Brother mindset.

I would like to have a chance to live with reordered expectations. I would rather that my country be seen as the rich, beloved brother than the rich and piggish one. If there's a heart beating in the United States that really disagrees, I've yet to meet it. We are, by nature, a generous people. Just about every American I know who has traveled abroad and taken the time to have genuine conversations with citizens of other countries has encountered the question, as I have, "Why isn't your country as nice as *you* are?" I wish I knew. Maybe we're distracted by our attachment to convenience; maybe we believe the ads that tell us that material things are the key to happiness; or maybe we're too frightened to question those who routinely define our national interest for us in terms of corporate profits. Then, too, millions of Americans are so strapped by the task of keeping their kids fed and a roof over their heads that it's impossible for them to consider much of anything beyond that. But ultimately the answer must be that as a nation, we just haven't yet demanded generosity of ourselves.

But we could, and we know it. Our country possesses the resources to bring solar technology, energy independence, and sustainable living to our planet. Even in the simple realm of humanitarian assistance, the United Nations estimates that $13 billion above current levels of aid would provide everyone in the world (including the hungry within our own borders) with basic health and nutrition. Collectively, Americans and Europeans spend $17 billion a year on *pet* food. We could do much more than just feed the family of mankind as well as our cats and dogs; we could assist that family in acquiring the basic skills and tools it needs to feed itself, while maintaining the natural resources on which all life depends. Real generosity involves not only making a gift but also giving up something, and on both scores we're well situated to be the most generous nation on earth.

We like to say we already are, and it's true that American people give of their own minute proportion of the country's wealth to help victims of disasters far and wide. Our children collect pennies to buy rain forests one cubic inch at a time, but this is a widow's mite, not a national tithe. Our government's spending on foreign aid has plummeted over the last twenty years, to levels that are—to put it bluntly—the stingiest among all developed nations'. In the year 2000, according to the Organization for Economic Cooperation and Development, the United States allocated just .1 percent of its gross national product to foreign aid—or about one dime for every hundred dollars in its treasury—whereas Canada, Japan, Austria, Australia, and Germany each contributed two to three times that much. Other countries gave even more, some as much as ten times the amount we do; they view this as a contribution to the world's stability and their own peace. But our country takes a different approach to generosity: Our tradition is to forgive debt in exchange for a strategic military base, an indentured economy, or mineral rights. We offer the hungry our magic seeds, genetically altered so the recipients must also buy our pesticides, while their sturdy native seed banks die out. At Fat Brother's house the domestic help might now and then slip out the back door with a plate of food for a neighbor, but for the record the household gives virtually nothing away. Even now, in what may be the most critical moment of our history, I fear that we seem to be telling the world we are not merciful so much as we are mighty.

In our darkest hours we may find comfort in the age-old slogan from the resistance movement, declaring that we shall not be moved. But we need to finish that sentence. Moved *from where?* Are we anchoring to the best of what we've believed in, throughout our history, or merely to an angry new mode of self-preservation? The American moral high ground can't possibly be an isolated mountaintop from which we refuse to learn anything at all to protect ourselves from monstrous losses. It is critical to distinguish here between innocence and naïveté: The innocent do not deserve to be violated, but only the naive refuse to think about the origins of violence. A nation that seems to believe so powerfully in retaliation cannot flatly refuse to look at the world in terms of cause and effect. The rage and fury of this world have not notably lashed out at Canada (the nation that takes best care of its citizens), or Finland (the most literate), or Brazil or Costa Rica (among the most biodiverse). Neither have they tried to strike down our redwood forests or our fields of waving grain. Striving to cut us most deeply, they felled the towers that seemed to claim we buy and sell the world.

We don't own the world, as it turns out. Flight attendants and bankers, mothers and sons were ripped from us as proof, and thousands of families must now spend whole lifetimes reassembling themselves after shattering loss. The rest of us have lowered our flags in grief on their behalf. I believe we could do the same for the 35,600 of the world's children who also died on September 11 from conditions of starvation, and extend our hearts to the fathers and mothers who lost them.

This seems a reasonable time to search our souls for some corner where humility resides. Our nation behaves in some ways that bring joy to the world, and in others that make people angry. Not all of those people are heartless enough to kill us for it, or fanatical enough to die in the effort, but some inevitably will be—more and more, as desperation spreads. Wars of endless retaliation kill not only people but also the systems that grow food, deliver clean water, and heal the sick; they destroy beauty, they extinguish species, they increase desperation.

I wish our national anthem were not the one about the bombs bursting in air, but the one about purple mountain majesties and amber waves of grain. It's easier to sing and closer to the heart of what we really have to sing about. A land as broad and green as ours demands of us thanksgiving and a certain breadth of spirit. It invites us to invest our hearts most deeply in invulnerable majesties that can never be brought down in a stroke of anger. If we can agree on anything in difficult times, it must be that we have the resources to behave more generously than we do, and that we are brave enough to rise from the ashes of loss as better citizens of the world than we have ever been. We've inherited the grace of the Grand Canyon, the mystery of the Everglades, the fertility of an Iowa plain—we could crown this good with brotherhood. What a vast inheritance for our children *that* would be, if we were to become a nation humble before our rich birthright, whose graciousness makes us beloved.

Knowing Our Place

I have places where all my stories begin.

One is a log cabin in a deep, wooded hollow at the end of Walker Mountain. This stoic little log house leans noticeably uphill, just as half the tobacco barns do in this rural part of southern Appalachia, where even gravity seems to have fled for better work in the city. Our cabin was built of chestnut logs in the late 1930s, when the American chestnut blight ran roughshod through every forest from Maine to Alabama, felling mammoth trees more extravagantly than the crosscut saw. Those of us who'll never get to see the spreading crown of an American chestnut have come to understand this blight as one of the great natural tragedies in our continent's history. But the pragmatic homesteaders who lived in this hollow at that time simply looked up and saw a godsend. They harnessed their mule and dragged the fallen soldiers down off the mountain to build their home.

Now it's mine. Between May and August, my family and I happily settle our lives inside its knobby, listing walls. We pace the floorboards of its porch while rain pummels the tin roof and slides off the steeply pitched eaves in a limpid sheet. I love this rain; my soul hankers for it. Through a curtain of it I watch the tulip poplars grow. When it stops, I listen to the woodblock concerto of dripping leaves and the first indignant Carolina wrens reclaiming their damp territories. Then come the wood thrushes, heartbreakers, with their minor-keyed harmonies as resonant as poetry. A narrow beam of sun files between the steep mountains, and butterflies traverse this column of light, from top to bottom to top again, like fish in a tall aquarium. My daughters hazard the damp grass to go hunt box turtles and crayfish, or climb into the barn loft to inhale the scent of decades-old tobacco. That particular dusty sweetness, among all other odors that exist, invokes the most reliable nostalgia for my own childhood; I'm slightly envious and have half a mind to run after the girls with my own stick for poking into crawdad holes. But mostly I am glad to watch them claim my own best secrets for themselves.

On a given day I may walk the half mile down our hollow to the mailbox, hail our neighbors, and exchange a farmer's evaluation of the weather (*terrible;* it truly is always either too wet or too dry in these marginal tobacco bottoms). I'll hear news of a house mysteriously put up for sale, a dog on the loose, or a memorable yard sale. My neighbors use the diphthong-rich vowels of the hill accent that was my own first language. My great-grandfather grew up in the next valley over from this one, but I didn't even know that I had returned to my ancestral home when I first came to visit. After I met, fell in love with, and married the man who was working this land, and agreed to share his home as he also shares mine in a distant place, I learned that I have close relatives buried all through these hollows. Unaccustomed as I am to encountering others with my unusual surname, I was startled to hear neighbors in this valley say, "Why, used to be you couldn't hardly walk around here without stepping on a Kingsolver." Something I can never explain, or even fully understand, pulled me back here.

Now I am mostly known around these parts by whichever of my relatives the older people still remember (one of them, my grandfather's uncle, was a physician who, in the early 1900s, attended nearly every birth in this county requiring a doctor's presence). Or else I'm known as the gal married to that young fella that fixed up the old Smyth cabin. We are suspected of being hard up (the cabin is quite small and rustic for a family of four) or a little deranged; neither alternative prevents our being sociably and heartily welcomed. I am nowhere more at home than here, among spare economies and extravagant yard sales glinting with jewel-toned canning jars.

But even so, I love to keep to our hollow. Hard up or deranged I may be, but I know my place, and sometimes I go for days with no worldly exchanges beyond my walk to the mailbox and a regular evening visit on our favorite neighbor's porch swing. Otherwise I'm content to listen for the communiqués of pileated woodpeckers, who stay hidden deep in the woods but hammer elaborately back and forth on their hollow trees like the talking drummers of Africa. Sometimes I stand on the porch and just stare, transfixed, at a mountainside that offers up more shades of green than a dictionary has words. Or else I step out with a hand trowel to tend the few relics of Mrs. Smyth's garden that have survived her: a June apple, a straggling, etiolated choir of August lilies nearly shaded out by the encroaching woods, and one heroic wisteria that has climbed hundreds of feet into the trees. I try to imagine the life of this woman who grew corn on a steeper slope than most people would be willing to climb on foot, and who still, at day's end, needed to plant her August lilies.

I take walks in the woods, I hang out our laundry, I read stories to my younger child, I hike down the hollow to a sunnier spot where I look after the garden that feeds us. And most of all, I write. I work in a rocking chair on the porch, or at a small blue desk facing the window. I write a good deal by hand, on paper, which—I somehow can't ever forget—is made from the macerated hearts of fallen trees.

The rest of the year, from school's opening day in autumn till its joyful release in May, I work at a computer on a broad oak desk by a different window, where the view is very different but also remarkable. In this house, which my predecessors constructed not from trees (which are scarce in the desert Southwest) but of sun-baked mud (which is not), we nestle into what's called in this region a *bosque*—that is, a narrow riparian woodland stitched like a green ribbon through the pink and tan quilt of the Arizona desert. The dominant trees are mesquite and cottonwood, with their contrasting personalities: the former swarthy with a Napoleonic stature and confidence, the latter tall and apprehensive, trembling at the first rumor of wind. Along with Mexican elder, buttonwillow, and bamboo, the mesquites and cottonwoods grow densely along a creek, creating a shady green glen that is stretched long and thin. Picture the rich Nile valley crossing the Saharan sands, and you will understand the fecundity of this place. Picture the air hose connecting a diver's lips to the oxygen tank, and you will begin to grasp the urgency. A riparian woodland, if it remains unbroken, provides a corridor through which a horde of fierce or delicate creatures may prowl, flutter, swim, or hop from the mountains down through the desert and back again. Many that follow this path—willow flycatchers, Apache trout—can live nowhere else on earth. An ill-placed dam, well, ranch, or subdivision could permanently end the existence of their kind.

I tread lightly here, with my heart in my throat, like a kid who's stumbled onto the great forbidden presence (maybe sex, maybe an orchestra rehearsal) of a more mature world. If I breathe, they'll know I'm here. From the window of my study I bear witness to a small, tunnelish clearing in the woods, shaded by overarching mesquite boughs and carpeted with wildflowers. Looming over this intimate foreground are mountains whose purple crowns rise to an altitude of nine thousand feet above the Tucson basin. In midwinter they often wear snow on their heads. In fall and early spring, blue-gray storms draw up into their canyons, throwing parts of the strange topography into high relief. Nearer at hand, deer and jackrabbits and javelina halt briefly to browse my clearing, then amble on up the corridor of forest. On insomniac nights I huddle in the small glow of my desk lamp, sometimes pausing the clicking of my keys to listen for great horned owls out there in the dark, or the ghostly, spine-chilling rasp of a barn owl on the hunt. By day, vermilion flycatchers and western tanagers flash their reds and yellows in the top of my tall window, snagging my attention whenever they dance into the part of my eyesight where color vision begins. A roadrunner drops from a tree to the windowsill, dashes across the window's full length, drops to the ground, and moves on, every single day, running this course as smoothly as a toy train on a track. White-winged doves feed and fledge their broods outside just inches from my desk, oblivious to my labors, preoccupied with their own.

One day not long ago I had to pull myself out of my writerly trance, having become aware of a presence over my left shoulder. I turned my head slowly to meet the gaze of an adolescent bobcat at my window. Whether he meant to be the first to read the story on my computer screen or was lured in by his own reflection in the quirky afternoon light, I can't say. I can tell you, though, that I looked straight into bronze-colored bobcat eyes and held my breath, for longer than I knew I could. After two moments (his and mine) that were surely not equal—for a predator must often pass hours without an eyeblink, while a human can grow restless inside ten seconds—we broke eye contact. He turned and minced away languidly, tail end flicking, for all the world a *cat*. I presume that he returned to the routine conjectures and risks and remembered scents that make up his bobcat-life, and I returned to mine, mostly. But some part of my brain drifted after him for the rest of the day, stalking the taste of dove, examining a predator's patience from the inside.

It's a grand distraction, this window of mine. "Beauty and grace are performed," writes Annie Dillard, "whether or not we will or sense them. The least we can do is try to be there." I agree, and tend to work where the light is good. This window is *the world* opening onto *me*. I find *I* don't look out so much as *it* pours in.

What I mean to say is, I have come to depend on these places where I live and work. I've grown accustomed to looking up from the page and letting my eyes relax on a landscape upon which no human artifact intrudes. No steel, pavement, or streetlights, no architecture lovely or otherwise, no works of public art or private enterprise—no hominid agenda. I consider myself lucky beyond words to be able to go to work every morning with something like a wilderness at my elbow. In the way of so-called worldly things, I can't seem to muster a desire for cellular phones or cable TV or to drive anything flashier than a dirt-colored sedan older than the combined ages of my children. My tastes are much more extreme: I want wood-thrush poetry. I want mountains.

It would not be quite right to say I *have* these things. The places where I write aren't actually mine. In some file drawer we do have mortgages and deeds, pieces of paper (made of dead trees—mostly pine, I should think), which satisfy me in the same way that the wren yammering his territorial song from my rain gutter has satisfied himself that all is right in *his* world. I have my ostensible claim, but the truth is, these places own *me:* They hold my history, my passions, and my capacity for honest work. I find I do my best thinking when I am looking out over a clean plank of planet earth. Evidently I need this starting point—the world as it appeared before people bent it to their myriad plans—from which to begin dreaming up my own myriad, imaginary hominid agendas.

And that is exactly what I do: I create imagined lives. I write about people, mostly, and the things they contrive to do for, against, or with one another. I write about the likes of liberty, equality, and world peace, on an extremely domestic scale. I don't necessarily write about wilderness in general or about these two places that I happen to love in particular. Several summers ago on the cabin porch, surrounded by summertime yard sales and tobacco auctions, I wrote about *Africa,* for heaven's sake. I wrote long and hard and well until I ended each day panting and exhilarated, like a marathon runner. I wrote about a faraway place that I once knew well, long ago, and I have visited more recently on research trips, and whose history and particulars I read about in books until I dreamed in the language of elephants. I didn't need to *be* in Africa as I wrote that book; I needed only to be someplace where I could think straight, remember, and properly invent. I needed the blessed emptiness of mind that comes from birdsong and dripping trees. I needed to sleep at night in a square box made of chestnut trees who died of natural causes.

It is widely rumored, and also true, that I wrote my first novel in a closet. Before I get all rapturous and carried away here, I had better admit to that. The house was tiny, I was up late at night typing while another person slept, and there just wasn't any other place for me to go but that closet. The circumstances were extreme. And if I have to—if the Furies should take my freedom or my sight—I'll go back to writing in the dark. Fish gotta swim, birds gotta fly, writers will go to stupefying lengths to get the infernal roar of words out of their skulls and onto paper. Probably I've already tempted fate by announcing that I need to look upon wilderness in order to write. (I can hear those Furies sharpening their knives now, clucking, *Which shall it be, dearie? Penury or cataracts?*) Let me back up and say that I am breathless with gratitude for the collisions of choice and luck that have resulted in my being able to work under the full-on gaze of mountains and animate beauty. It's a privilege to live any part of one's life in proximity to nature. It is a privilege, apparently, even to know that nature is out there at all. In the summer of 1996 human habitation on earth made a subtle, uncelebrated passage from being mostly rural to being mostly urban. More than half of all humans now live in cities. The natural habitat of our species, then, officially, is steel, pavement, streetlights, architecture, and enterprise—the hominid agenda.

With all due respect for the wondrous ways people have invented to amuse themselves and one another on paved surfaces, I find that this exodus from the land makes me unspeakably sad. I think of the children who will never know, intuitively, that a flower is a plant's way of making love, or what *silence* sounds like, or that trees breathe out what we breathe in. I think of the astonished neighbor children who huddled around my husband in his tiny backyard garden, in the city where he lived years ago, clapping their hands to their mouths in pure dismay at seeing him pull *carrots* from the *ground.* (Ever the thoughtful teacher, he explained about fruits and roots and asked, "What other foods do you think might grow in the ground?" They knit their brows, conferred, and offered brightly, "Spaghetti?") I wonder what it will mean for people to forget that food, like rain, is not a product but a process. I wonder how they will imagine the infinite when they have never seen how the stars fill a dark night sky. I wonder how I can explain why a wood-thrush song makes my chest hurt to a populace for whom wood is a construction material and thrush is a tongue disease.

What we lose in our great human exodus from the land is a rooted sense, as deep and intangible as religious faith, of why we need to hold on to the wild and beautiful places that once surrounded us. We seem to succumb so easily to the prevailing human tendency to pave such places over, build subdivisions upon them, and name them The Willows, or Peregrine's Roost, or Elk Meadows, after whatever it was that got killed there. Apparently it's hard for us humans to doubt, even for a minute, that this program of plunking down our edifices at regular intervals over the entire landmass of planet earth is overall a good idea. To attempt to slow or change the program is a tall order.

Barry Lopez writes that if we hope to succeed in the endeavor of protecting natures other than our own, "it will require that we reimagine our lives. . . . It will require of many of us a humanity we've not yet mustered, and a grace we were not aware we desired until we had tasted it."

And yet no endeavor could be more crucial at this moment. Protecting the land that once provided us with our genesis may turn out to be the only real story there is for us. The land *still* provides our genesis, however we might like to forget that our food comes from dank, muddy earth, that the oxygen in our lungs was recently inside a leaf, and that every newspaper or book we may pick up (including this one, ultimately, though recycled) is made from the hearts of trees that died for the sake of our imagined lives. What you hold in your hands right now, beneath these words, is consecrated air and time and sunlight and, first of all, a place. Whether we are leaving it or coming into it, it's *here* that matters, it is place. Whether we understand where we are or don't, that is the story: To be *here* or not to be. Storytelling is as old as our need to remember where the water is, where the best food grows, where we find our courage for the hunt. It's as persistent as our desire to teach our children how to live in this place that we have known longer than they have. Our greatest and smallest explanations for ourselves grow from place, as surely as carrots grow in the dirt. I'm presuming to tell you something that I could not prove rationally but instead feel as a religious faith. I can't believe otherwise.

A world is looking over my shoulder as I write these words; my censors are bobcats and mountains. I have a place from which to tell my stories. So do you, I expect. We sing the song of our home because we are animals, and an animal is no better or wiser or safer than its habitat and its food chain. Among the greatest of all gifts is to know our place.

Oh, how can I say this: People *need* wild places. Whether or not we think we do, we *do*. We need to be able to taste grace and know once again that we desire it. We need to experience a landscape that is timeless, whose agenda moves at the pace of speciation and glaciers. To be surrounded by a singing, mating, howling commotion of other species, all of which love their lives as much as we do ours, and none of which could possibly care less about our economic status or our running day calendar. Wildness puts us in our place. It reminds us that our plans are small and somewhat absurd. It reminds us why, in those cases in which our plans might influence many future generations, we ought to choose carefully. Looking out on a clean plank of planet earth, we can get shaken right down to the bone by the bronze-eyed possibility of lives that are not our own.

The Patience of a Saint

Written with Steven Hopp

When I was nine years old I jumped across the Mississippi. My family had sought out its headwaters in Itasca State Park, Minnesota, where a special trail showed the way for adventurers with this feat in mind. I took my leap reverently. My parents made sure I understood that this modest stream I'd taken in stride was actually one of the earth's great corridors, the dominion of paddleboats and Huck Finn, a prime mover of flood, fertility, and commerce across our land.

However much we may

long to re-create the landmark events of our own childhoods for our children, water passes on. One can't—as Heraclitus put it—step into the same river twice. Nowadays when my family sets out for a lesson in river, we often drive southeast from Tucson to a narrow, meandering cottonwood forest where the kids may attempt to vault the San Pedro. They've done it often and sometimes don't even get very wet. Where its headwaters cross from Mexico into Arizona, this river is barely three feet across. As it runs north across a hundred miles of desert with a scant but persistent flow, it rarely gets much wider. In the scheme of human commerce it's an unimpressive trickle. Mostly it's a sparkling anomaly, a novelty for us here—a thread of blue-green relief for sunstruck eyes.

In the heat of late April the modest saint invites us down from the blazing desert into a willowy tunnel of cool shade, birdsong, and the velvet-brown scent of riverbank. We take unhurried hikes there whenever we can, reading the dappled script of animal tracks and the driftwood history of flood and drought embedded in the steep banks. The sight of a vermilion flycatcher leaves us breathless every time—he's not just a bird, he's a punctuation mark on the air, printed in red ink, read out loud as a gasp.

The kids dance barefoot between sandbars, believing they have found the Secret Garden. For the space of an afternoon we're sheltered from the prickly reality of the desert in which we live. Most human visitors to the San Pedro appreciate it for about the same reasons people value gold: It sparkles, and it's rare.

From a resident's point of view, though, the price of gold couldn't touch this family home. For the water umbel spreading delicate roots in a lucid pool, the leopard frog peering out through a veil of duckweed, the brush-prowling ocelot, and the bright-feathered birds that must cross this hostile expanse of land or find a living on it, the San Pedro is a corridor of unparalleled importance. Nearly half the river's hundred miles, and fifty-eight thousand acres of the surrounding corridor, have been protected since 1988 as the San Pedro Riparian National Conservation area. The Nature Conservancy has named it one of the nation's "Last Great Places."

To jump across this river with the right measure of reverence requires an animal frame of mind. Eighty-two species of mammals—a community unmatched anywhere north of the tropics—inhabit this valley. Hiding out here as well are 43 kinds of reptiles and amphibians, including the endangered Huachuca leopard frog, a bizarre critter that calls (as if he knew it was a big, harsh desert out there) from underwater. The San Pedro also harbors the richest, densest, and most diverse inland bird population in the United States—385 species. It's one of the last nesting sites for willow flycatchers and western yellow-billed cuckoos; green kingfishers breed nowhere else in the country. For millions of migratory birds traveling from winter food in Central America to their breeding grounds in the northern U.S. and Canada, there is one reliable passage, on which their lives depend. Just this one.

I lead my children down its banks in the hope that they'll come to recognize in the San Pedro the might and consequence of that splendid word *river*. Never mind that Huck Finn wouldn't have troubled himself to spit across it. As our girls stoop at the edge of a riffle, peering into the clear, fast water, my husband and I talk to them about heroic navigational feats undertaken not by paddle and steam but by feathered wing.

Our Tucson-born children are more accustomed to ephemeral desert streams that roar briefly after a storm, leaving behind bleached, stony channels that stay dry for weeks or months until the next good rain. "This one never, ever dries up. Wow," the elder observes. "There could be fish living in there." Her little sister, meanwhile, hurls herself toward the sandy shallows, crying, "Clothes off!"

This may or may not be reverence, but most children are good at the animal frame of mind.

This place is one of the blessings I count when I brace myself to consider a dearly beloved and threatened world, and stake my heart onto pieces of what's left of it. The pulse of a whole continent beats in this thinly drawn vein, and I'm called to put my hand in it and listen. Within the leafy protectorate of the conservation area, most of the river is flanked by a trail, and much of it we've walked, in sections as long as a Saturday and a small child's legs. We hike northward with the flow, guessing the river's intentions as it braids into sandbars and shallows. Sometimes it nearly disappears, but stands of young cottonwoods testify to a permanent flow just under the surface. It's more troubling to find a grove of old trees with no young ones at their feet; this means the water table, depleted by nearby wells, has dropped too far for saplings to take hold. If it continues to drop, eventually even the grandfather cottonwoods that constitute the backbone of this ecosystem won't be able to find sufficient moisture to sustain themselves.

When we stop to listen for the yellow-breasted chat in a thicket of Mexican elders, or scan the water ahead for the greenish glint of a kingfisher's wings, we have in the back of our minds, always, the health of this river. We visit the San Pedro as one visits a beloved, elderly relative: We don't talk about the inevitable, but we think about it a lot.

Rivers like this were once common in the Southwest, with permanent flow supporting long corridors of cottonwood-willow gallery forest that netted the body of the Sonoran Desert like veins. Now, in a land bled dry by agriculture and population growth, only 5 percent of the original forest remains. Riparian species have become concentrated in disparate fragments of creekside habitat. The species list is still impressive, but among many orders of animals and plants it's not as long as the roll call of those that have perished quietly. Of the fourteen species of fish that were once native here, twelve are now extinct. Beavers used to dam the rivers into strings of marshy pools, keeping the water table high, but they were hunted out long ago. It's no surprise to a desert hiker anymore to top a hill and look down on a ghost parade of giant, leafless cottonwoods snaking through the valley below, dying in place, marking a watercourse that has gone and won't ever come back.

Yet the San Pedro somehow perseveres. Scattered along its hundred miles are artifacts from countless human encroachments, beginning with that of the Clovis people; they settled this valley eleven thousand years ago, when woolly elephants roamed our continent and San Pedro with his Pearly Gates was far from anyone's mind. Hunting these marshlands with flint arrows, the Clovis settlers established the most prosperous North American population that archaeologists have found from that era. The Mogollon and Hohokam later built on their foundations, then themselves gave way to the Apaches, who controlled the corridor until the Spanish arrived. After Coronado's exploratory party beat down this path in 1540, the San Pedro became the point of entry for Spanish settlement of North America. Three centuries later pioneers from the East rushed to lay claim to every water source, especially this river, because claiming water was the only strategy that ensured survival here on an eastern bureaucrat's idea of a homestead allotment. Forty acres of desert will support a herd of pack rats but not a subsistence farm. However, with forty-acre plots strung like beads up the watercourse, and huge tracts of adjacent desert left largely unoccupied, each rancher with a water hole had an effective claim on many thousands of acres of rangeland.

Now the family cemeteries of homesteaders are tucked back among the trees, along with the crumbling adobe ghosts of boomtowns whose economies lived and died by mining. A roaring mill on the riverbank once pulverized copper ore hauled in by mule from nearby Tombstone. The river provided water for processing and a handy place to dump the toxic byproducts. Now we explore these ruins gingerly, cautioning the kids not to climb on fragile walls or impale themselves on scrap iron as they try to spy pottery shards and other interesting junk. It seems strange to say it, but I am comforted by all this faint and crumbling evidence of human civilizations that have risen and fallen before us. People come and go, as plans begun so modestly inevitably burgeon and bluster until the land beneath our feet finally fails to support our big ideas. It's not the end of the world, at least insofar as the frogs and fishes are concerned. Here among the green willows I am always tempted to see the remnants as reassurance: I want to believe that our own century's harsh claims on the river will someday ease away from it just as gracefully, so that our legacy will be absorbed with all the others' into San Pedro's patient embrace.

But since the day Coronado first guided his horses through tall sacaton grass in this valley, skirting a marshy river then hundreds of feet across, we've changed the face of the land almost unimaginably. Now the river is channelized between steep banks, and so reduced that at one point, near the farming town of Benson, the entire flow runs through an irrigation sluice. It's a dramatic lesson in ecology, a calculus effected largely by the subtraction of just one native from the river community and the addition of a single outsider: the beaver and the free-range cow.

Ranchers are testy about their claim here; it's a tough enough life they've inherited, without city-bred environmentalists challenging their rights to graze and irrigate. But now, as the Sunbelt booms, farmers and environmentalists find their voices equally drowned out by a new, louder demand from urban consumers. Historically, most of Arizona's water has gone to agriculture—80 percent at present, with the remainder divided between industrial and municipal use. The state's Department of Water Resources predicts, however, that municipal consumption will more than double over the next half century. Much of the population growth is expected in southeastern Arizona, where the San Pedro flows. The Fort Huachuca army base and the growing city of Sierra Vista flank the San Pedro, drinking up groundwater from an underground depression that diverts and depletes the same subterranean channel that supports the river and its many forms of life. As the trees die, fingers point in every direction, and well they might: On a map this corridor resembles a patchwork quilt of ranchland, nature preserve, townships, and government grazing leases. A handful of residents and environmentally minded Arizonans who love this river for what it is, rather than for what they can take from it, are now working against the clock to set aside pieces of the water table and crucial tributaries, aiming to place them out of reach of human depletion.

But a river doesn't flow in pieces. Migratory routes can't skip over private property, and a fish has little use for a river that runs 90 percent of the time. Recently the San Pedro's troubles came to the attention of the Commission for Environmental Cooperation, newly created by the North American Free Trade Agreement (NAFTA). A study of the crucial migratory corridor connecting Mexico, the United States, and Canada was an appropriate first undertaking for this trinational body charged with fostering cooperative environmental responsibility. The findings were unequivocal: At the present rate of consumption, the report indicated, human occupancy will dry up the San Pedro in a matter of decades. Capping municipal growth here, limiting irrigation, and closing Fort Huachuca could significantly extend the river's life, but as the study conceded, these would be costly measures in human terms.

How can the San Pedro's case be argued in a human tongue whose every word for "value" is tied to the gold standard of human prosperity? I feel frustrated in trying even to frame the question, for it occurs to me the question is this: If life must be a race to use up everything we have, who exactly will win that race?

The land offers other kinds of answers. This blue-green slice of life, fiercely bounded on every side, will continue to try to persist for all it is worth. To themselves and one another, the little lives in this watery sphere mean the world. But their delicate finned and feathered hopes are at this moment being weighed against thirsts beyond their ken.

This knowledge makes it harder for us, each time, to return to our beloved river, but impossible also to stay away. Our family goes back in every season now, not just for cool relief on hot days but to witness the autumn migration of hawks and stand under trembling cottonwoods as they cast golden leaves from their white-skinned arms. In winter we kneel at the base of a bare old netleaf hackberry to place our palms against its bark and feel the mysterious rows of raised bumps that stipple the trunk like letters in a manuscript written in Braille. What can we read there? How long before the pages of this book peel and dry to dust?

At the end of our Saturday hike we leave the scent of mud and moss and find a path through dense elders into the sparser shade of mesquites and then, finally, the end of the trail. In the parking lot we hesitate, seeing that a vermilion flycatcher has taken a position in a branch above our car. He sits like a sentry with his puffed-out chest, sallying out suddenly to snap a mosquito from the air, then returning to his post. For several minutes we watch him from a distance, unwilling to intrude on the territory he's claimed so convincingly.

Eventually he flits off, startled by a car that has bumped down the gravel off-ramp from the highway: a station wagon from Ohio. An elderly tourist climbs out, stretches his limbs, and takes a look around.

Then he glances at us, and at the parking lot that seems to suggest he has arrived at some destination. "What is this place, anyway?" he asks.

"The San Pedro River," we answer, more or less reverently.

"A river," he repeats slowly, casting a dubious eye on the cactus-studded hills around us. "How big is it?"

We glance at one another, abashed for our river. Evidently this is a question we don't know how to answer. "About three feet across" can't possibly be right.

As big as life, then. Just that.

Seeing Scarlet

Written with Steven Hopp

Picture a scarlet macaw: a fierce, full meter of royal red feathers head to tail, a soldier's rainbow-colored epaulets, a skeptic's eye staring out from a naked white face, a beak that takes no prisoners.

Now examine the background of your mental image; probably it's a zoo or a pet shop, metal bars and people chanting about Polly and crackers, maybe even pirates, and not a trace of the truth of this bird's natural life. How does it perch or forage or speak among its kind without the demeaning mannerisms of captivity? How does it look in flight against a blue sky? Few birds that inhabit the cultural imagination of Americans are so familiar and yet so poorly known.

As biologists who have increasingly turned our attention toward the preservation of biodiversity, we are both interested in and wary of animals as symbols. If we could name the passion that kept pushing us through Costa Rica in our rented jeep, on roads unfit for tourism or good sense, we would have called it, maybe, macaw expiation. Some sort of penance for a lifetime of seeing this magnificent animal robbed of its grace. We wanted to get to know this bird on its own terms.

As we climbed into the Talamanca Highlands on a pitted, serpentine highway, the forest veiled the view ahead but always promised something around the next bend. We were two days south of San José, in a land where birds lived up to the extravagance of their names: purple-throated mountain gems, long-tailed silky flycatchers, scintillant hummingbirds. At dawn we'd witnessed the red-green fireworks of a resplendent quetzal as he burst from his nest cavity, trailing his tail-feather streamers. But there'd been no trace of scarlet yet, save for the scarlet-thighed dacnis (yes, just his thighs—not his feet or lower legs). Having navigated through an eerie morning mist in an elfin cloud forest, we found ourselves at noon among apple orchards on slopes so steep as to make the trees seem flung there instead of planted. All of it was wondrous, but we'd not yet seen a footprint of the beast we'd come here tracking.

Then a bend in the road revealed a tiny adobe school, its bare-dirt yard buzzing with activity. The Escuela del Sol Feliz took us by surprise in such a remote place, though in Costa Rica, where children matter more than an army, the sturdiest shoes are made in small sizes, and every tiny hamlet has at least a one-room school. This one had turned its charges outdoors for the day in their white and navy uniforms, so the schoolyard seemed to wave with neat nautical flags. The children, holding tins of paint and standing on crates and boxes, were busy painting a mural on the school's stucco face: humpbacked but mostly four-legged cows loafing beneath round green trees festooned with round red apples, fantastic jungles dangling with monkeys and sloths. In the center, oversize and unmistakable, was a scarlet macaw. The children's portrait of their environment was a study in homeland, combining important features of both real and imaginary landscapes; while their macaw surely had more dignity than Long John Silver's, he was still a fantasy. All of these children had picked apples and driven their family cows across the road, and some may have seen the monkeys they depicted in their mural, but not one, probably, had ever laid eyes on a macaw.

Ara macao was once everywhere in Costa Rica—in the lowlands at least, on both the Pacific and Atlantic coasts—but in recent decades it has been pushed into a handful of isolated refuges as distant as legend from the School of the Happy Sun. Its celebrity in the school's mural cheered us because it seemed a kind of testimonial to its importance in the country's iconography, and to the sporadic but growing effort to teach children here to take their natural heritage to heart. We'd come here in search of both things: the scarlet macaw and some manifestation of hope for its persistence in the wild.

Our destination was the Corcovado National Park on the Osa Peninsula, where roughly a thousand scarlet macaws constitute the most viable Central American population of this globally endangered bird. The Osa is one of two large Costa Rican peninsulas extending into the Pacific; both are biologically rich, with huge protected areas and sparse human settlements. Corcovado, about one tenth the size of Long Island, is the richest preserve in a country known for its biodiversity: Its bird count is nearly 400 species, and its 140 mammals include all 6 species of cats and all 4 monkeys found in Central America. It boasts nearly twice as many tree species as the United States and Canada combined. The park was established by executive decree in 1975, but its boundaries weren't finalized until many years later, after its hundreds of unofficial residents had been relocated. Hardest to find were the gold prospectors—who had a talent for vanishing into the forest—and the remnant feral livestock, which disappeared gradually with the help of jaguars.

For us, Corcovado would be the end of a road that was growing less navigable by the minute as we ventured farther out onto the peninsula. Our overnight destination was Bosque del Cabo, a private nature lodge at the peninsula's southern tip. Our guidebook had promised that we'd cross seven small rivers on the way, but we didn't realize we'd have to do it without the benefit of bridges. At the bank of the first one we plunged right in with our jeep, fingers crossed, cheerfully encouraged by a farmer in rubber boots who was leading his mule through the water ahead of us.

"This will be worth it," Steven insisted when we reached the slightly more treacherous-looking second river. There was no bridge in sight, and no evidence that one had ever existed, although a sign advised *Puente en mal estado*—"Bridge in a bad state." Yes indeed. The code of Costa Rican signs is a language of magnificently polite understatement; earlier in the trip we had been informed by a notice posted on a trail leading up a live volcano, "Esteemed hiker, a person can sometimes be killed here by flying rocks."

Over the river safe and sound, with the Golfo Dulce a steady blue horizon on our left, we rattled on southward through small fincas under the gaze of zebu cattle with their worldly wattles and huge downcast ears. Between farms the road was shaded by unmanicured woodlots, oil-palm groves, and the startling monoculture of orchard-row forests planted for pulp. The dark little feathered forms of seedeaters and grassquits lined the top wires of the fences like intermittent commas in a run-on sentence. To give our jostled bones and jeep a break, we stopped often; any bird was a good enough excuse. A dark funnel cloud swirling above a field turned out to be a vast swarm of turkey- and black vultures. With our binoculars we scanned the vortex down to its primogenitor: a dead cow, offering itself up for direct recycling back into the food chain. Most of the peninsula's airborne scavengers, it seemed, had just arrived for dinner. Angling for position near the carcass, two king vultures flapped their regal black and white wings and rainbow-colored heads at each other. "Wow, amazing, gorgeous!" we muttered reverently, gawking through our binoculars, setting new highs in vulture admiration.

At dusk, with seven rivers behind us, we pulled into the mile-long driveway of Bosque del Cabo under a darkening canopy of rain forest. Although the road tunneled between steep, muddy shoulders, we could smell the ocean beyond. Our headlight beam caught a crab in the road, dead center. We slid to a stop and scrambled out for a closer look at this palm-sized thing. A kid with a box of Crayolas couldn't have done better: bright purple shell, red-orange legs, marigold-colored spots at the base of the eye stalks. We dubbed this beauty "resplendent scarlet-thighed crab" and nudged it out of the road, only to encounter more just like it almost immediately. Suddenly we were seriously outnumbered. Barbara surrendered all dignity and walked ahead of the jeep in a crouch, waving her arms, but as crabherd she was fighting a losing battle against a mile-deep swarm. These land crabs migrate mysteriously in huge throngs between ocean and forest, and on this moonlit night they caught us in a pulsing sea of red that refused to part. They danced across the slick double track of their flattened fellows, left by other drivers ahead of us. In our many trips together we've rarely traveled a longer, slower, *crunchier* mile than that one.

We slept that night in a thatched *palapa,* lulled by the deep heartbeat of the Pacific surf against the cliff below us. At first light we woke to the booming exchanges of howler monkeys roaring out their ritual "Here I am!" to position their groups for a morning of undisturbed foraging. We sat on our little porch watching a coatimundi already poking his long snout into the pineapple patch. A group of chestnut-mandibled toucans sallied into a palm, bouncing among the fronds; no macaws, though we were in their range now. We walked out to meet this astonishing place, prepared for anything except the troop of spider monkeys that hurled sticks from the boughs and leapt down at us, hanging from their prehensile tails in a Yankee-go-home bungee-jumping display. Retreating toward our lodge, we heard a parrotish squawk in the treetops that we recognized from pet shops. Was it a macaw?

"Sí, guacamayos," we were assured by a gardener whom we found shaking his head over the raided pineapple patch. Yes, he'd been seeing macaws lately, he said, usually in pairs, *"practicando a casarse"*—"practicing to be married." This was April, the beginning of nesting season. Following their species' courtship rituals, the macaw pairs would settle into tree cavities, always situated more than a hundred feet off the ground, to lay and hatch their two-egg clutches. The young stay with their parents for up to two years, during which time the adults do no more nesting until after these young have dispersed. This combination of specialized habitat and slow reproduction makes macaws especially vulnerable to an assembly of threats. The ravages of aerial pesticide spraying have lately diminished, as banana companies have left the country or switched to oil-palm production, but deforestation remains a

phenomenal peril. Of the macaws' original Costa Rican habitat, only 20 percent still stands, all of it now protected. In addition to the Osa population, there are some 330 birds in the Carara Biological Reserve to the north, and others survive in scattered pockets from southern Mexico into Amazonian Brazil.

Dire habitat loss has become the norm for tropical species, but macaws and parrots are further doomed by their own charm. Such beauty doesn't come cheap: A poacher who captures a young scarlet macaw can sell it into the pet trade for up to $400 U.S. (The fine for being caught is about $325.) Since 1990, when the nearby town of Golfito was allowed to reduce taxes on goods passing through its port, employment in the import trade has grown, and poaching has noticeably declined. Farther north, however, in the economically undeveloped Carara region, the activity is still ubiquitous. Many conservationists feel that their best hope is to introduce alternative sources of income for the poachers while educating their children about poaching's long-term tradeoffs—which could include the extinction of a national emblem before they're old enough to become adept at climbing hundred-foot trees. During our trip we spoke with several educators whose programs in the schools are aimed specifically at developing a family conscience about stealing baby parrots and macaws from their nest holes. Reordering children's attitudes toward threatened species may eventually influence their families, so the thinking goes, even within a culture that has traditionally allowed these birds to be harvested with no more moral qualms than a hungry coatimundi brings to a pineapple patch.

An organization named Zoo Ave goes a step further, by rehabilitating birds recovered from poachers or from captivity and reintroducing them into the wild. So far the group has released nineteen birds into the forest on the eastern shore of Golfo Dulce, far enough from Corcovado that the populations should remain genetically distinct. Of these birds, eleven are known to have survived. Zoo Ave's goal is to establish a population of a dozen or so breeding pairs in the area near Rainbow Lodge, adjacent to the newly protected Piedras Blancas National Park. Given that the total number of breeding macaw pairs in Central America is probably less than a hundred, every new nest cavity lined and filled with two white eggs is cause for celebration.

"El que quiera azul celeste, que le cueste," the Costa Ricans say—"If you want the blue sky, the price is high." The mix of hope and fatalism in this *dicho* speaks perfectly of the macaw's fierce love of freedom and its touching vulnerability. We stood on a cliff near our *palapa* above the ocean, scanning, hoping for a glimpse of scarlet that wasn't there. Today we would complete our determined pilgrimage to Corcovado, where we would see them flying against the blue sky, or we would not. On a trip like this, you revise your hopes: If we saw even one free bird, we decided, that would be enough. We prepared to push on to the road's end at Carate, gateway to the Corcovado forest, home to the country's last great breeding population of scarlet macaws.

Carate, though it appears on the map, is not a town; it's a building. Mayor Morales's ramshackle *pulpería* serves the southwestern quadrant of the peninsula as the area's singular hub of commerce: He'll arrange delivery-truck passage back out to Puerto Jiménez, buy the gold you've mined, watch your vehicle for a small fee while you hike, or simply offer a theoretical rest room among the trees out back. Indoors, suspended by wires from the ceiling, is a dazzling assortment of bottles, driftwood, birds' nests, car parts, and other miscellany—the very definition of "flotsam and jetsam," provided you can tell what floated in and what was jettisoned.

Above the main counter dangles the crown jewel of the collection: a mammalian vertebra of a size generally seen only in museums. Under this Damocles bone we purchased a soda and plotted our strategy for finding macaws. Outside on benches under a tree, we learned from the *pulpería* regulars that Corcovado did not exactly offer the user-friendliness we were accustomed to in a national park. There were no roads in, no hiking trails, no wooden-signboard maps declaring that "you are here." How do we get in? we asked. You walk, and watch out for snakes, we were told. It's a thick jungle; where's the best walking? On the beach.

While we chatted, a pet spider monkey sidled up to Barbara on her bench. Steven focused the camera. Nearby, a barefoot girl was watching intently.

"Is he friendly?" Barbara asked in Spanish.

The girl grinned broadly. *"Muerde"*—"He bites."

Steven snapped the photo we now call "Interspecific Primate Grimace."

The steep gray beach offered rugged access to the park. The surf pounded hard on our left as we hiked, and to our right the wall of jungle rose sharply up a rocky slope. A series of streams poured down the rocks from the jungle into the Pacific. At the forest's edge the towering trees were branchless trunks for their lower hundred feet or so. We began to hear, from this sparse, lofty canopy, the sound of macaws—not the loud, familiar croak but a low, conversational grumbling among small foraging groups. We jockeyed for a view, catching glimpses of monkeylike movement as the birds clambered around pulling fruit from clusters at the tips of branches. Macaws are seed predators, given to cracking the hearts of fruit seeds or nuts. High above the ground is where you'll see them, only and always, if you don't want a cage for a backdrop. Both the scarlet and the other Costa Rican macaw, the great green, require large tracts of mature trees for foraging, roosting, and nesting.

It's hard to believe that anything so large and red could hide so well in foliage, backlit by the tropical sky, but these birds did. We squinted, wondering if this was it—the view we'd been waiting for. Suddenly a pair launched like rockets into the air. With powerful, rapid wingbeats and tail feathers splayed like fingers they swooped into a neighboring tree and disappeared again uncannily against the branches. We waited. Soon another pair, then groups of threes and fives, began trading places from tree to tree. Their white masks and scarlet shoulders flashed in the sun. A grand game of musical trees seemed to be in progress as we walked up the beach counting the birds that dived between branches.

All afternoon we walked crook-necked and openmouthed with awe. If these creatures are doomed, they don't act that way: *El que quiera azul celeste, que le cueste,* but who could buy or possess such avian magnificence against the blue sky? We stopped counting at fifty. We'd have settled for just one—that was what we thought we had come for—but we stayed through the change of tide and nearly till sunset because of the way they perched and foraged and spoke among themselves, without a care for a human's expectations. What held us there was the show of pure, defiant survival: this audacious thing with feathers, this hope.

Setting Free the Crabs

At the undulating line where the waves licked the sand on Sanibel Island, our three pairs of human footprints wove a long, sinuous path behind us. Littoral zone: no-man's-land, a place of intertidal danger for some forms of life and of blissful escape for others. The deliberate, monotonous call and response of the waves—assail, retreat—could have held me here forever in a sunlight that felt languid as warm honey on my skin. So we moved in a trance, my mother, my daughter, and I, the few sandblasted clamshells and knotty whelks we had gathered clacking together in the bag that hung carelessly from my fingertips. Our practiced beachcombers' eyes remained on high alert, though, and eventually my daughter's eye caught the first true find of our day: a little horse conch, flame orange, faceted, perfect as a jewel. *Treasure.*

My daughter wanted to take it home, I know. She turned it over, already awed like any lottery winner by the stroke of sudden wealth and the rapid reordering of the mind that tells itself, *Yes! You did deserve this.*

And then her face fell. "Uh-oh," she said. "Already taken."

"Oh, shoot," my mother said. "Is it alive?" There are laws, on Sanibel, about taking live creatures from the ocean.

"Well, not the conch—that's gone. But a hermit crab's in the shell."

Two small white claws protruded from the opening. The sluggish gastropod that had been architect and builder of this magnificent orange edifice had already died—probably yesterday, judging from the condition of the shell—but as any house hunter can tell you, no home this gorgeous stands empty for long. A squatter crab had moved in.

"Oh, they don't care if you take *those,*" my mother reassured her. "There are thousands of hermit crabs on this beach."

She was right, of course, though I could not help thinking, There are thousands of us on this beach, too—at what point do we become expendable? But I said nothing, because I had nothing sure to say, and anyway I was more interested in hearing how my daughter would respond. I decided to watch my leggy, passionate ten-year-old walk into the jaws of this dilemma by herself.

She looked up, uncertain. "But it's a living creature, Grandmama. We can't kill it just because we want a shell for our collection."

My mother, like every grandmother, wants her grandchildren to have the sun, the moon, and the stars, all tucked into a box with a bright red bow. If my daughter really wanted this shell, Grandmama was going to give her an out. "Well," she said, summoning remarkable creativity, "can't we find it another shell?"

My daughter pondered this. She knows, as I do, that a hermit crab won't give up its shell just because you want it. It will hold on. It will relinquish a claw or a head, or whatever else you manage to pull off, rather than come out. Were we going to take this thing home and set out an array of alternatives in front of it, as if it were a hapless shopper who'd won a dazzling spree? Some hermit crabs, the bigger ones with reddish claws, are game for a certain amount of terrestrial adventure, but this one wasn't that kind. Away from the littoral zone, this tiny life would give up its ghost within a few hours. I know this, I'm ashamed to say, from experience. So I waited, as did my husband, who had jogged up to join us, wondering what our little life-and-death huddle was all about.

My daughter looked at the creature in her hand for a long time and then said firmly, "No. We can't kill it."

"Anyway, it has the best shell on this whole beach," Steven said, quick to nail a few planks of support to her decision lest it should wobble. "It deserves to keep it."

So we handed it over to him, and he tossed it far out into the surf, to brood out there however a crustacean mind may brood upon a catastrophe narrowly escaped in the cradle of a human child's hand.

I have tried to teach my children to love nature as my parents taught that reverence to me—through example, proximity, and plenty of field guides and age-appropriate biology books. As long as I've been conscious of my thoughts, I've considered myself a lover of nature. Only when I was old enough to have fallen in and out of love with other things and people did I begin to understand that there were different kinds of love. There is the sort I think of as maternal—both selfless and wholly giving—the point of which is to help some other life do as well as possible even outside your presence, and hopefully to survive beyond you. Even if the object of your affection moves, say, to New Zealand, and you know you're never going to see it again, you will still love it, and love it fiercely. You'll send it food, money, anything.

Then there is a less selfless, more possessive form of passion. This may be what most of us felt for our first high school flame: a desperate need to be near, to observe, to show it off, to have and to hold.

I understand that I waver between these kinds of love when I throw my heart to nature. I cherish the wild things in my backyard, but I also love that I get to be near them. I need to live *somewhere,* I reason; the house was already built when I got here, so I will be a responsible steward of the place and take it under my wing. It's easy when that stewardship coincides with my own needs, but not so much fun when these programs collide and I am forced to feel more like what I really am: a colonist on occupied territory. When the cute wild things charge down the fence around my garden and bury their faces in my watermelons, they're not cute anymore; then they're the *un*cutest damn ugly things I've ever laid eyes on. I count to ten. I don't shoot them. We are working this out.

And we are working it out at the level of species, as we arrive slowly at our new understanding that we are wiping nature off our map before we have ever even had the chance to get to know it well. As recently as my grandfather's generation, critters and varmints were unquestionable enemies—reasonably enough, I suppose, considering that however much my kids enjoy our watermelons, my grandfather's kids literally survived on the crops he grew. Back then, the going assumption on such creatures was simple: You couldn't shoot them fast enough out of your fields and orchards, and if you could eat the corpses that fell, so much the better. There will always be more passenger pigeons where those came from.

When humans decide to work our will, we are so tragically efficient. Now that we've used up all the prairie, we've taken to burning the rain forests to clear pasture for fast-food beef, slashing and burning a new plot of it the size of Tennessee each year, without much of anyplace left to go after that. In the meantime, and largely as a result, the rate of species extinctions has reached astounding new highs; many scientists predict we will lose about a quarter of the world's wildlife over the next two decades. "We ignore these losses at our peril," warned Al Gore in his brilliant book *Earth in the Balance.* "They're like the proverbial miners' canaries, silent alarms whose message in this case is that living species of animals and plants are now vanishing around the world one thousand times faster than at any time in the past 65 million years." What we're witnessing now is the most catastrophic extinction event since the dinosaurs died—it looks like Rome is burning. And plenty of people are fiddling as it burns: In November 2000 exactly half of the voters in this country opposed the man who wrote the words I just quoted. But the other half voted *for* him, I remind myself. Right now, other frightening imperatives have distracted us so far from the program of benevolence toward our planet that it seems we might just try to burn the whole world for fuel to keep ourselves guarded and cozy. But that is *not* the expressed will of our people. Most of us do understand, when we can calm down and think clearly, that whether we are at peace or at war, the lives that hang in the balance are not just ours but the millions more that create the support system and biological context for humanity. More and more of us are listening for the silent alarm, stopping in our tracks, wishing to salvage the parts of this earth we haven't yet wrecked.

Even in our best-intended efforts, though, it's hard to sort out goodwill from self-interest. We want every square inch of our national parks to be accessible by paved road and private automobile, with rest rooms ever handy. We work lots harder to save the panda than to rescue the snail darter, presumably because the latter is such a plain little fish that we don't much care whether or not our children will ever get to see one cavorting in a zoo. I do not begrudge the lovely pandas one penny of their save-the-panda money, heaven knows they need it, but I worry that our bias toward saving "charismatic megafauna" (as a friend of mine calls them) begets a misguided strategy. If we believe in putting women and children in the lifeboat first, we should look harder at ecosystems to see what's at the bottom of their replication, cleanup, and maintenance—the crucial domestic labor of a planet, the grunt work that keeps everything else alive. That is: soil microbes, keystone predators, marine invertebrates, pollinating insects, and phytoplankton, oh my. The day I see a Soil Microbe Beanie Baby, I'll know we're getting somewhere.

It seems impossible that humans could view the world with less immediate self-interest, and yet it *isn't*. The first people to inhabit North America arrived here without a biblical mandate to go forth and have dominion over the fish of the sea, the fowl of the air, and every thing that creepeth upon the earth. Those first Americans had different stories, which allowed them to take an utterly different approach: Their culture accepted the sovereignty of the animate land that fed and sheltered them, and it held that certain mountains and valleys were too sacred even to visit. Even then, humans were sometimes too successful for the earth's good; the prevailing theory on the extinction of the mastodons and other large Ice Age mammals in North America is that people came along and ate them, every one. But the point is, they didn't have a Manifest Destiny that claimed this destruction as righteous. The religion of their descendants, at least, precludes such smugness.

Those of us who come from a different tradition may find it hard to grasp as sacred the notion of inaccessible wilderness persisting all on its own, unlooked upon by human eyes, preserved simply for the sake of its grandfathered membership in the biodiverse club of wild habitats on this planet. For most modern Americans, that amounts to the sound of one hand clapping. We'll preserve the wilds, sure, but we still want to own them somehow, and take home a snapshot as proof.

Some friends unknowingly proposed a dilemma about nature-love in a story they told me of visiting Cancún years ago. It was still a relatively sleepy fishing town then, surrounded by paradise, poised on the edge of discovery and prompt destruction. These friends are of a certain age and were far ahead of their time in the manner of appreciating nature. Over the course of their lives they have dedicated a great deal of their energy to conservation.

"We saw what was coming to Cancún," they said. "We actually saw the bulldozers starting to clear it. So we saved what we could out of that jungle. We have orchids growing in our greenhouse that we collected from there."

I admired their enterprise and empathized with their heartbreak at seeing delicate, rare lives crushed. And yet if it had been my choice to make, I think I'd have felt uneasy at the prospect of profiting in any way—even just aesthetically—from the destruction of a sacred place. Maybe I'm wrong about this, or maybe there really is no right way to look at it, but my heart tells me it's better to grieve the whole loss than to save a handful of orchids. Better to devote oneself to anger and bereavement, to confront the real possibility that soon there will be nowhere left to go, anywhere, to see an orchid in the wild, than to derive a single iota of pleasure from these small, doomed relics of a home that's forever gone. Anger and bereavement, throughout history, have provided the engine for relentless struggles for change. In a greenhouse these orchids will flourish awhile and then, after a few years or many, die. A jungle is a form of eternal life, as ephemeral and enduring as the concept of love or mystery. It cannot be collected.

More recently, modern science has settled this question by working with governments to place strict limits on the collection and transportation of native species, especially endangered ones. Although the market for contraband exotics still persists (and this perverse appetite regularly precipitates horror stories such as the one about the parrots smuggled from Mexico inside automobile hubcaps), the net effect of the government limits has been to discourage private possession of morally unownable things. A zoo is many steps up from a private collection, at least in its modern form as a park where the animals are given more space to roam and more species-appropriate habitats than the humans who must walk down narrow paved trails to see them. Most modern zoos have signed on to the proposition that they are in business not just to let kids have a gawk at a giraffe or an elephant, but also to join in the worldwide effort to spare giraffes and elephants from extinction on their home ground. From fund raising and reproduction programs to the sponsorship of significant research, most zoos are more about animal advocacy (and increasingly, habitat advocacy) than about possession. And perhaps most important of all, they offer the only opportunity that most modern children (and adults) will ever have to get any real sense of what biodiversity is. The American Zoo and Aquarium Association (AZA) requires its member institutions to manage captive animals in a manner that furthers their conservation, and half the U.S. population passes each year through AZA facilities—that's more people than attend all professional football, basketball, and baseball games combined. This is a significant contribution to our nation's education, reaching far beyond the population that actively supports environmental projects. Once individuals have experienced "lion," not just with their eyes during a TV nature show but with their ears, nose, and the little hairs that stand up on the back of your neck when a lion stares you down, they can be expected to share the world with lions in a different way—a way, we can hope, that will be more protective of the animals' right to occupy their own place. The first steps toward stewardship are awareness, appreciation, and the selfish desire to have the things around for our kids to see. Presumably the *un*selfish motives will follow as we wise up.

Meanwhile, we grapple with what it really means to love animals. My husband, an ornithologist who studies bird populations, was once amazed, in a little, out-of-the-way pet shop, to see an Indian hill mynah on display in a cage. He asked if there was a captive breeding population of these birds—a possibility that seemed unlikely. The man in the store said no, the mynah had been captured in the wild in India and brought here to be sold as a pet. My husband was shocked to hear that; these birds were already known to be declining, though this was some years before their capture and sale became strictly illegal. He asked how the pet-store owner could justify selling a bird that was in danger of being extirpated from the wild.

"We're keeping it safe," the man explained without a twinge of remorse. "Somebody will take very good care of it."

"But you've taken it from the wild. It's gone from the breeding population," my husband protested.

"But it's right here, still alive," the man replied.

"Yes, but you've essentially killed it. Even if there were a mate for it somewhere, they probably wouldn't reproduce, and that'd be a dead end anyway. Genetically speaking, this bird is dead."

The pet-shop fellow looked at his bird, which must have seemed to him very much alive, and insisted, "It's extremely dangerous for these birds in the wild. By keeping this one as a pet, we've saved its life."

Both men restated their arguments a few times until it was clear they had reached an impasse. My husband left the man and the bird that day, but he has never stopped thinking about this semantic deadlock over what it means to "save the animals." For all of us whose first biology lesson was Noah's ark, it is hard to unlearn the fallacy that sparing just a few of anything can provide some sort of salvation. It takes a basic knowledge of population genetics to understand exactly why a breeding population of a certain size, in a healthy habitat, is necessary for the continuation of a species. Low genetic variation, inbreeding, and lethal genes all mean that when a population gets down to the last two of a kind, they might as well be just one; their species is doomed. Certainly a single bird in a cage, separated from its habitat and its species, is done for. Orchids without the mystery of their forest are not what they were; likewise, an Indian hill mynah removed from its Indian hills is nothing but an object of beauty. No longer in its own sense a living thing, it has become a possession.

The trick here is to distinguish between caring about the good of a species and caring about an individual creature. These two things can actually run at cross purposes. One animal lover, for example, may be putting out seed to attract birds and help them through the winter, even as the animal lover next door is nurturing a cat bent on carrying out a methodical campaign of genocide (or rather, avicide) at the bird feeder. This is not to suggest that it's wrong to love a cat or a dog, or to sell or buy pets, or to lobby for animal rights in the form of better treatment for cats, dogs, veal calves, or lobsters about to be put into boiling pots, but these concerns do not make an environmental casc. They make a spiritual case, and animal-rights activists are practicing a form of religion, not environmental science. I like to think that the world is plenty large enough for both science and religion, and usually the two mesh well. But sometimes they may confuse or contradict each other. Certainly my own relationships with the animals in my life are absurdly complex: Some I love, some I eat, and the scraps left over from the ones I eat, I feed to the ones I love. (Is there a song about that?) But as I try to sort this out, I find that when I must choose, my heart always comes down on the side of biodiversity.

A famous conflict between these interests arose when the

Nature Conservancy undertook to preserve the very last few hundred acres of native Hawaiian rain forest. This tract is a fragile fairyland of endemic ferns and orchids that were being rooted to shreds by feral pigs. Anything native to Hawaii has no defenses against ground predators, simply because these ecosystems evolved without them; thus, nene geese don't run or fly when humans approach, and native birds are helpless against the mongoose-come-latelies that eat their eggs. For the flora, the problem is pigs: The Polynesians brought them over in their canoes for food (they would later be replaced by larger pigs brought in European ships), and some escaped to the wild, where their descendants now destroy every root in their path. The Nature Conservancy faced an animal lover's painful dilemma. The extremely difficult terrain and the caginess of the wild hogs made it impossible to take them alive; to save the endangered forest some pigs would have to be killed. Enter, then, the People for the Ethical Treatment of Animals (PETA), who set up a remonstration. The Conservancy staff argued that sparing a few dozen pigs would cost thousands of other animal and plant lives and extinguish their kinds forever. They also pointed out that the pigs had come to Hawaii in the first place under a human contract, as a food item. No matter, said PETA; the chain of pig death ends here. The two groups have reached some compromises, but the ideological conflict remains interesting.

I applaud any religion that devotes itself to protecting life; I applaud it right up to—but stopping short of—the point where protecting one life-form brings an unintended holocaust upon others that are being overlooked. In this contest between a handful of pigs and thousands of native birds, insects, and plants, neither side could fairly say it was simply advocating *life*. It had become necessary to make a choice between systems—restoring a natural one versus upholding an increasingly damaged one.

For the sake of informed choices, I took a trip. I walked in that magical forest, by special invitation, so I might carry out a story that would wring compassion from people far away who would never get to see its wonders. The story is a heartbreaker, so I did the best I could. I could already see the ghosts of the place; it was that near death, and that willfully alive. White mists rose through the curved spines of blue-green fern trees. A single scarlet bird with a sad, down-curved bill spoke its name, *iiwi*, again and again, like the eulogy a child might sing for himself if every last relative had died of the plague. I want that place to *be,* forever. I will never step on that soft moss again, I don't want to leave any more footprints, but I would give anything for that scarlet iiwi to find a mate and produce two small eggs and a future of songs among those ferns. I felt sorrow for being human there and ached for the ignorance of my kind, who seem always to arrive in paradise thinking only of our next meal. For this bowl of lentils—a pork chop, a can of sliced pineapple rings—we sold our birthright to paradise and infected Hawaii with a plague on its native kind.

This story of pigs and forest is a tale about possession. A pig can be owned; an iiwi can't. Pigs are a human invention, as are cows, Chihuahuas, and house cats. Over thousands of years our ancestors transformed wild things into entirely new species that they named food, work, or companionship. These beasts are alive, as surely as the yeast that makes our bread is alive, but they are animals only by our definition, not by nature's. They have no natural habitat. However much we may love them, or not, they are our things, like our houses and vehicles. When cats or dogs or pigs go wild, the effect on nature is something like what would happen if our useful yeast were to transform itself into an Ebola virus: It begins a cascade of deaths one after another, extending far beyond the reach of what we ourselves have bulldozed or killed. Scientists who study this destruction have estimated, for example, that domesticated cats in North America kill as many as four million songbirds *every day.* (The millions of feral cats out there—those that have left human habitation and are fed by no one but themselves—add many more deaths to this toll.) These animals are a living extension of our possession. There must be limits, somewhere, to the human footprint on this earth. When the whole of the world is reduced to nothing but human product, we will have lost the map that can show us how we got here, and can offer our spirits an answer when we ask why. Surely we are capable of declaring sacred some quarters that we dare not enter or possess.

A sad loss recently befell my friends, the orchid growers who witnessed the sad destruction of Cancún many years ago: The large, forested lot next to their home was cleared for development. They had been assured, from the time they moved into their house, that the beautiful piece of wild land abutting them was not for sale. But everything has its price, it seems, and now when I visit them we sit on the porch facing away from the absurdly huge, modern house that was built next door, right up to the edge of its lot on every side, and though we don't speak of it, we are mourning. Perhaps there really is no such thing as saving the wilderness next door for our own enjoyment. Enjoyment goes only with the enjoyers, who will be the death of this place—of every place. People love the woods but can't abide the mosquitoes, so we spray insecticide from airplanes, which ends up killing not just mosquitoes (and the encephalitis germ we dread) but also monarch butterflies, ladybugs, lacewings, and the birds and lizards that eat the poisoned ants.

My daughter, a few years after she surrendered the world's best shell to that hermit crab, did a science-fair project on the aerial mosquito spray of choice, malathion, and its effects on life beyond mosquitoes. She discovered that at unbelievably minute concentrations it still causes the tiny microorganisms in our wetlands to swim in desperate circles and then die. This zooplankton—

uncharismatic though it may be—is the staff of life, the stuff that supports the tiny fish, which support the bigger fish, which are eaten by raccoons and bears and herons and people and bald eagles. The toxin kills the bugs that pester you, and another million creatures that you've never thought about or even noticed. From an insect's point of view, let's face it, the obliteration of all to punish the perceived crimes of an infinitesimal percentage amounts to precisely the horror that we humans have named, in our own world, ethnic cleansing.

I don't know if the average human mind can open wide enough to think of it that way. Last night I slapped a mosquito that was drinking from my arm and then stared awhile at the little splat, feeling mildly avenged at the sight of my foe's blood until I realized, of course, that the blood was my own. Oh, what a tangled web we weave when first we practice to do the right thing! We take care of ourselves, we destroy; we don't take care of ourselves, we destroy. Mosquitoes, I have been told, are important pollinators in the Arctic. So, good, they have their place in the grand scheme, and I'll vote against aerial spraying on behalf of everything else that goes before the fall, but it's taking me some time to get to that emotional plane where I can love a mosquito. It may in fact require more than a few lifetimes' remove from the varmint-killing ethic whence I arose. My generation has taken historic steps toward appreciating nature, setting aside more parklands, and enjoying them in greater numbers than any before it. But if we are going to hold on to this place in any form that includes genuine wilderness, we will have to become the kind of people who can imagine a faraway, magical place like the Arctic National Wildlife Refuge—and all the oil beneath it—and declare that it is not ours to own because it already owns itself. It's going to demand the most selfless kind of love to do right by what we cherish, and to give it the protection to flourish outside our possessive embrace.

Maybe that step begins with giving up ownership of the most beautiful shell on the beach, not simply to save the life of a homely, ordinary crab, but as an exercise in resisting the hunger to possess all things bright and beautiful. It can begin when a ten-year-old mind senses the sovereignty of living worlds apart from her own, so that a perfect shell may be—must be—thrown back into the sea. We humans have fallen far from the grace we once had, when we could look on every mountain with fear and reverence, but we have also crept slowly back from the depths, when we needed to have our names carved on every mountaintop and a passenger pigeon in every pot. We seem mostly to be moving in some kind of right direction, if only we aren't too late. I hope my own mistakes will serve as a benchmark for my children, to show them how life accumulates its wisdom and moves on.

A Forest's Last Stand

Xmul. X'pujil. Once you learn to pronounce the *X* as a "Shh . . . ," the place-names of the Mayas sound like so many whispered secrets. So does the Mayan language that is still spoken, with quiet ubiquity, in the Yucatán. Along rural roadsides, where fathers and sons walk in early light to the milpas, you can hear it. At the Merida market where women sit and lean their heads together behind stacks of tomatoes and *chaya* leaves, this language of secrets is passed along.

Leading south from the colonial city of Merida to the ruins of ancient Uxmal is an old road that rises into dry hills of farms and woodlands. This was the road we chose. As Steven drove, I navigated, using a map that showed a Mesoamerican culture's famous antiquities while somehow neglecting to mention that the culture itself was still completely alive. This was Mayan countryside. Nearly every little town had an *X* to its name, and every woman who walked along the roadside had on the Mayan dress, a lace-trimmed white cotton tunic brilliantly embroidered at the bodice and hem. All of the dresses were different, like eye-popping snowflakes, and they obviously weren't put on for tourists—we were by this time well out of tourist terrain. The women wore them when they did their marketing, laundering, and garden work, and even, as I saw once, when they fed the family hogs; miraculously, the dresses always seemed to remain dazzlingly white. To my eye, this was magical realism.

Our journey's end lay much farther to the south, in the humid forests that touch the Guatemalan border, but I tend to travel toward destinations the same way I look up words in the dictionary, getting sidetracked by every possible item of interest along the way. So we made a detour to inspect one of the notable antiquities on our map: Uxmal. Older by centuries than the aggressively heroic pyramids at Chichén Itzá, Uxmal's structures are just as tall but somehow less high-and-mighty. (They are also far less frequently visited by tourists, because they aren't handily reached from Cancún.) The Pyramid of the Magician is round-shouldered and delicate—if the latter word can reasonably be used of a pyramid. Also the plaza at Uxmal is less grandly paved, more mossy underfoot than Chichén Itzá's. The soft ground swallowed the sounds of our steps as we walked through the enormous, silent city. Everywhere we looked, the facades were etched with turtles, monkeys, and jaguars, and the staircases guarded by feather-headed serpents. Living iguanas the size of small alligators perched on the cornerstones, glaring at us in a good imitation of the glowering stone heads above them. The limestone rain gods at Uxmal have looked down their huge, up-curled noses for nearly two thousand years, but the iguanas have less patience with the enterprise; they're inclined to roll off their posts and undulate across the weathered steps. Winding paths lead out from the central plaza through the forest to other clearings and buildings: temples, ball courts, stories carved in stone. The fringe of surrounding jungle hides dozens more structures that have been left unrestored, consigned to crumble quietly under blankets of vine and strangler fig, keeping their secrets to themselves. As we walked the forest path, a light rain began to darken the gods' stone pates, imperceptibly dissolving their limestone, carrying off another small measure of history.

As we left the modern settlement that surrounds Uxmal, we recalled the counsel of friends in Merida not to head south into the sparsely populated state of Campeche without a full tank of gas. Adventurous but not foolish, we backtracked to the nearest PEMEX. Steven negotiated for *sin plomo* while I attacked the windshield, facing up to an omelette of Mexican insect life. I was mostly still lost in Uxmal's iguana dreams from the day before, but suddenly as I scraped at the windshield I found my attention snagged on a gigantic agrichemical-company ad painted on the building across the street. It showed a merry campesino dousing his corn with a backpack sprayer as huge green letters loomed in the sky above him: *Psst . . . Psst . . . There goes your security!*

Something must be getting lost here (or gained) in my translation, I thought; this was just too much truth in advertising. If Mexico is the NAFTA sister with the brightly embroidered dress and the hibiscus behind her ear, she is also the one whose reputation has been most tarnished by chemical dependency. The fields here are dumping grounds for DDT and virtually anything else ever deemed too toxic for U.S. consumers. The country's capital has our planet's most chronically poisoned air; the 1994 earthquake, in momentarily shaking traffic to a halt, afforded many of Mexico City's residents their first-ever sight of blue sky. The chemicals sign might just as well say *Psst . . . There goes your tourist dollar!* It's true that Mexico's siren song of beaches and margaritas calls out seductively to students on spring break, but North American travelers looking for nature unspoiled generally skip right over it to Costa Rica and points south.

In doing so, they fly directly over the place we were now setting out to look for on our full tank of gas, a putative gem of undefiled Mexico. Most people would be surprised to learn that the largest tropical forest on our continent stands on the southernmost reach of Yucatán and northern Guatemala.

By late afternoon we had reached its frontier, the Calakmul Biosphere Reserve. The tiny town of X'pujil ("Shh . . . pujil") guards a bend in the road and the ruins of Becán, a city even more ancient than Uxmal and even less visited by tourists. We stopped to walk Becán's tree-filled plazas and secret stone passageways in the company of no other humans at all, only birds. A turquoise-browed motmot peered down at us from a branch, its long tail clicking back and forth like the pendulum of a clock.

My map showed that the reserve pinched to a wasp-waist here at X'pujil. The wilderness broadens out as it stretches north to the Puuc Hills and far south into Belize and the Guatemalan highlands. In its southern reaches, this same forest shadows the ancient Mayan cities of Tikal and Uaxactun. The road we were on skirted the forest's edge. The sun had set, but the trees' upper branches were still lit like candles, aflame with birds. Keel-billed toucans hailed us from high overhead with huge beaks that looked freshly painted by an artist on a binge. In the treetops they threw back their heads and laughed their good-nights. An enormous lineated woodpecker's vermilion crest stood straight up as if frozen in fright—possibly by the news that his next of kin, the imperial and ivory-billed woodpeckers, have gone extinct. We rolled down our windows and breathed in rarefied steam. The boughs of a gumbo-limbo tree drooped low with roosting chachalacas, dark, chicken-size birds renowned for their remarkable singing style. But by now it was too late in the day for singing. Eyes in shining pairs blinked from the roadside: foxes, agoutis, maybe wild cats. The Calakmul Reserve is home to jaguarundis, ocelots, margays, pumas, and jaguars. It also conceals tapirs and opossums, turkeys colored like peacocks, orchids and bromeliads the size of turkeys, monkeys that hoot like owls, and owls with eyes in the back of their heads. Where the map shows a vast green emptiness, the land is alive.

That this wilderness still exists is something of an accident of geology. In this century the industries of deforestation moved south through Mexico like General Sherman, sweeping and burning it clear of its subtropical forests. The march came this far and then literally dried up. Even though the rains come heavy at times here, no streams at all cross the Yucatán's limestone surface; the same water scarcity that plagued the ancient Mayans has daunted modern ranchers in the region, preventing the successful development of this land for large-scale meat farming. So the Calakmul, for now, still belongs to jaguars and toucans.

Of course, birds and beasts alone have no power to save a Mexican forest. What the Mayans of old worshiped as gods, the modern Mayans tend to eat. Likewise the Chol, Tzeltal, and other groups fleeing Guatemalan repression and Mexican poverty. Some fifteen thousand refugees poured into this region in the mid-1990s. Their tradition of slash-and-burn farming demanded that they leave behind used-up cornfields every three years to clear new patches of forest. The Mexican government had designated the Calakmul forest a biosphere reserve in 1989, but signs posted to that effect tended to be read by the homeless as invitations to settle. *"Hooray,"* refugees must have exclaimed at the sight of the Forest Preserve signs. "Nobody's living here who will give us trouble."

And so Mexico's last great forest, having held its own against timber magnates and hamburger franchises, seemed doomed to fall one branch at a time to cookfires and corn patches. But because of an extraordinary program launched in 1991, it still stands. In the villages surrounding the Calakmul another whisper was going around, maybe a feather of hope for the place. That was what we were looking for.

In the village of Nueva Vida, or "New Life," Carmen Salgado waved happily from her gate and invited us into her backyard garden. We told her we'd been sent from the *consejo*. Just north of the ruins at X'pujil we'd found the small concrete-block building that housed the farmers' co-op office, and the kind folks there had directed us to Nueva Vida. They'd promised that here and in the region's other small villages we might find an intriguing update on the civilizations we had been admiring in postmortem condition to the north, where the great pyramids poked out of the Yucatán forests. Here and now, in a cooperative of thirty-six families, papaya and lime trees shaded thatched houses elegantly constructed of smooth wooden poles. I kept studying them until the connection registered: These high-peaked roofs perfectly echoed the shape of the vaulted ceilings we'd seen inside every Mayan ruin we had visited. The architecture had preserved its central elements for thousands of years.

But here was another story—the village of New Life was looking very much alive. Outside Carmen's high-peaked thatched house, in her sunny garden, I stepped carefully to avoid solid plantings of cilantro, lettuce, and *chaya*—which she explained was a high-protein leaf crop that had been grown in the area since ancient times. A vine she called nescafé curled its tendrils around the wire fence that contained her compost pile; from its beans she made a coffee substitute and protein-enriched bread. We walked from her back gate down the gravel path through the village center, where a lush community citrus orchard offered oranges and grapefruits. A turkey paused to eye us, then continued stalking the ground under the citrus trees with a fierce forager's eye, taking seriously his job as the DDT of a new generation.

No snaking backpack sprayers will *psssst* in this Garden of Eden. Carmen informed us in no uncertain terms that chemical pesticides and fertilizers are beyond the means of the subsistence farmers here—and what's more, they are learning not to want them. Instead, they demoralize pests with a concoction of soap, onions, and garlic. Their reliance on organic methods of pest control and soil amendment allows these farmers self-sufficiency, while also ensuring that their notoriously poor tropical soil will improve with each crop, rather than deteriorate.

Carmen's broad, handsome face lit up as she explained these things. Although she has had almost no formal education, she is astute, articulate, and comfortable with visitors, a natural spokesperson for Nueva Vida and its new program. She grew up in one place and another in the poorest parts of rural Monterrey, without family land or much hope until she came here. She was lucky: She arrived just as a new environmental appreciation was dawning over the Calakmul forest, and with it a new approach to its conservation. Everything depends on these villages immediately surrounding the forest preserve. Nueva Vida is one of the seventy-two *ejidos,* or cooperative farms, that ring the Calakmul Reserve in a protective belt, established by land grants assigned to groups of refugee families that otherwise, inevitably, would have consumed the forest from the inside out. The plan the reserve's managers came up with may seem contradictory to U.S. notions of wilderness preservation, but here in the land of the Maya it may just be the only right solution: Rather than fight a losing battle to keep people out, they would help them move *into* the forest. Recognizing that human habitation was an ancient and integral part of this ecosystem, the managers hoped that nature might best be preserved here by human residents who had a good enough reason to care for it. A boundary of settlements could buffer the forest against waves of outsiders' moving farther in. The program's goal was to encourage these farmers to shift their long-standing war against trees into a peaceful coexistence.

But having a land grant means staying in one place and learning to call it home, no small departure for the refugee populations of Nueva Vida and the other *ejidos,* who previously spent their lives using up land and moving on. The concept of composting may seem obvious enough to the sedentary, but for those with no cultural memory of standing still for more than three years, seeing soil improve and fruit trees grow is a kind of miracle. It's almost impossible to explain what a huge leap of faith is involved here, even in a citrus orchard. I was at first amused and then, as I began to understand, profoundly impressed by the enthusiasm Carmen and her fellow *ejidarios* displayed for their orchards and gardens and even their simple, beautifully functional composting toilets. Watching things grow, improving a piece of land—for historical refugee populations these are cultural accomplishments even more significant than learning to read and write or earning a degree. They embody a complete psychological transformation.

The transition has happened gradually, in daily lessons that come through patience and careful scrutiny. Don Domingo Hernández, an elder statesman in the neighboring collective of Valentín Gómez Faríaz, had a lot to tell us about that. He walked us out to his cornfield, where he was experimenting with soil-boosting cover crops, and gave us a lively lecture on the benefits of chemical-free agriculture: healthy soil microbes, nitrogen fixation, humus, conservation of moisture. Don Domingo tipped back his weathered cowboy hat, bent to scoop up a handful of black dirt, and held it out to me in his hand as reverently as any true believer might handle a relic of his faith. "Three years in this patch," he said, "and this is the best corn crop I've ever had. Next year will be even better."

The prime mover behind this change was not government charity but education provided by Pronatura, a Mexican conservation group, in concert with the U.S.-based Nature Conservancy and other private organizations. Working from a thatch-roofed office just north of X'pujil—which, like its demonstration garden, is open to visitors—a handful of agronomists and engineers offer ideas and technical advice; they are enormously respected by the region's farmers. Carmen was animated about this point. "Doña Norma came out from the *consejo* office and said to us, 'What are you wasting time for? Plant trees!' So we planted trees." This and dozens of other projects have given families like Carmen's a sense of belonging to their land, and a reason to stay. They have dug cisterns to catch rainwater from their roofs, following the instructions of a Pronatura engineer; they have also begun new beekeeping enterprises. A course in medicinal plants offered at the *consejo* office teaches women how to collect, process, and label all the remedies their families will need for infections and minor ailments. (The medicines are stored in plastic film canisters donated by conservation groups in the city.) After Hurricane Roxana devastated the southern Yucatán, when fevers and infections were scathing the peninsula, Carmen's collective had kilos of medicine on hand to donate to the relief effort.

"It's a much better life we have now," Carmen insists. "We were skeptical at first, and some still hold to the old ways. But really, the old way was that we ate rice and beans and drank coffee. The rice and coffee, we had to buy with cash. We have a healthier diet now with all the things we grow, and it's nicer, more interesting. Our kids like it better—that's how you know a change is going to stick." She gave my belly a glance, smiling at my incipient pregnancy. "For the kids, there is no going back; this is the life they will choose."

The stalwart tropical day had slipped away into evening as we were talking, and we now stepped outside Carmen's cool thatched house to watch the full moon rise. Orchids planted in tin cans bloomed pale and fragrant in the dusk. Normally they grow in the forest canopy, unseen by human eyes; Carmen called these her *huérfanas pobrecitas,* or "poor little orphan girls," because she'd salvaged them from trees that the men of a neighboring *ejido* had felled for lumber. Every collective includes arable fields and a parcel of forest land extending into the Calakmul Reserve, to be used as the cooperative sees fit. Some are cutting their trees, sustainably, in a managed forestry program, but increasingly, others are not cutting theirs at all. Carmen's group of women voted against clearing their twenty-five hectares for a cornfield, deciding that the parcel would be more valuable to them as it stood, since it provides flowers year-round for beekeeping as well as an inexhaustible apothecary. It's also a balm for the spirit. Carmen made us pause in our conversation to look at the moon, a perfect orange lantern cradled in the arms of a cecropia tree.

"Listen!" she commanded, her eyes bright. From the forest's edge a warm wind carried the scent of wild spices and the sweet call of a pygmy owl. Somewhere within the jungle nearby a blue-eyed jaguar crouched, searching the wind for signs of its age-old forest companion, the human animal.

Many kilometers from the bordering ring of villages, deep in the very heart of the reserve, the giant pyramids of the Calakmul ruins rest in permanent peace. At each turn of our journey we'd seen more remote, unvisited ruins, but this was literally the end of the road—the very edge of North America, beyond which no human residence or enterprise was to be found for a far cry. Our new friends from the *ejido* and *consejo* office had roused us in the early-morning darkness from the small thatched house on stilts where we'd spent the night, guiding us down the long, bumpy dirt road into the forest's heart with promises of the most dramatic sunrise of our lives. Now we groped our way by flashlight up deeply weathered steps to the top of the tallest pyramid. Mayan glyphs silently held their accounts beneath the industry of foraging ants. As the limestone softly crumbles, the forest retrieves it.

On a small platform atop a pyramid that was probably used for just such dramatic ceremonies in antiquity, our little group waited for the sunrise. I stood up, a little dizzy from the height—we were way above the treetops—and out of breath from the steep climb. I put my hand on my belly, where I carried the daughter whose name and gender I didn't yet know, and I whispered to my child internally: *Remember this with me. Once upon a time we were here, at the center of the world.* As far as I could see in every direction, a dark green sea of untouched forest rolled out to the whole, encircling horizon. In a lifetime—mine, anyway—one is given this blessing only rarely: the chance to stand on high ground, turn in every direction, and see absolutely not one single sign of visible humanity. This is how the world once was, without our outsize dreams and dominion. Nothing surrounded us but the dark embrace of trees, except where the predawn light touched the eroded stone face of another pyramid rising above the canopy. Our friends pointed out a bump on the southern horizon that they said was the pyramid of Mirador, on the Guatemalan border. From here to there, when the sun was just right, a person could flash signals with a mirror; and from Mirador, someone else could signal farther south to Tikal, and so on, to the edge of the Mayan world. We stood at its very center. Then, between one held breath and the next, the sun appeared to us, scarlet and full-skirted on the horizon.

Very suddenly we found ourselves surrounded not by eerie silence but by a wilderness of wake-up calls. A troop of howler monkeys began to stir in the treetops just below us, letting loose a loud, primordial bellow. Emerald battalions of parrots darted past in formation, flashing in the whitewashed light.

Then came the chachalacas, the chickenlike birds we'd seen the previous day, whose call, I had been promised, I would never forget. *"Shh . . . !"* our friends said, *"Escuche,"* and we listened, but I didn't hear it at all. And then I did: a barely audible chorus in the far distance, Cha-*chalac?* Quietly, distantly, their neighbors answered back, *Cha*-chalac! The more I listened, the more plainly I could hear how they followed the call-and-response rhythm of a gospel choir: cha-*chalac? Cha*-chalac! They stirred one another to voice in increasing numbers to announce their revelation. This forest, I began to understand with a chill, was entirely filled with chachalacas. The birds themselves don't move, but their song does as they awaken one another each morning, their dawn chorale moving through the whole jungle in a vast oratory wave. The rising tide of their gospel song raced toward us, growing louder, louder and faster: Cha-*Chalac? CHA*-CHALAC! CHA-CHALAC! *Glory hallelujah!* The song came from everywhere at once, a musical roar like water, and then like water it divided, passing around us as a rush of singing, and then it receded and fell away—*Cha-chalac!*—toward the southern horizon. Finally it faded out of earshot.

None of us spoke. I imagined this wave of hallelujah traveling all the way to Guatemala and beyond, on down to the southern edge of the jungle, where the trees once again gave way to roads and cornfields, billboards and gas stations. But we were still deep inside a green, crowded world where parrots and monkeys were not isolated survivors but citizens of a population. It was a city of animals here, as surely as each mute temple stood for a city of people who had once carved their reverence for animals in stone and climbed up to greet the dawn.

Of course, they aren't gone, the Mayans. On carved slabs of stone they left us clear pictures of their world, with man and beast facing off nose to nose in a thousand configurations: warrior and monkey; jaguar and emperor. The Mayans' ways and reverences have endured like stone, altered through seasons of sun and rain. Now in our latter days, iguanas scowl at tourists, and farmers may raise up great clouds of death on the vermin, but sometimes another story can root itself and take hold. In some quarters, farmers named Carmen and Don Domingo rule, in a reign that allows no poison and holds its breath for the moon and smiles at the sweet nightsong of an owl. Human and beast together may persist in this place, as they have always done, since the days when God was a feather-headed serpent.

Called Out

Written with Steven Hopp

The spring of 1998 was the Halley's Comet of desert wildflower years. While nearly everyone else on the planet was cursing the soggy consequences of El Niño's downpours, here in southern Arizona we were cheering for the show: Our desert hills and valleys were colorized in wild schemes of maroon, indigo, tangerine, and some hues that Crayola hasn't named yet. Our mountains wore mantles of yellow brittlebush on their rocky shoulders, as fully transformed as eastern forests in their colorful autumn foliage. Abandoned cotton fields—flat, salinized ground long since left for dead—rose again, wearing brocade. Even highway medians were so crowded with lupines and poppies that they looked like the seed-packet promises come true: that every one came up. For weeks, each day's walk to the mailbox became a botanical treasure hunt, as our attention caught first on new colors, then on whole new species in this terrain we thought we had already cataloged.

The first warm days of March appear to call out a kind of miracle here: the explosion of nearly half our desert's flowering species, all stirred suddenly into a brief cycle of bloom and death. Actually, though, the call begins subtly, much earlier, with winter rains and gradually climbing temperatures. The intensity of the floral outcome varies a great deal from one spring to another; that much is obvious to anyone who ventures outdoors at the right time of year and pays attention. But even couch potatoes could not have missed the fact that 1998 was special: Full-color wildflower photos made the front page of every major newspaper in the Southwest.

Our friends from other climes couldn't quite make out what the fuss was about. Many people aren't aware that the desert blooms at all, even in a normal year, and few would guess how much effort we devote to waiting and prognosticating. "Is this something like Punxsutawney Phil on Groundhog Day?" asked a friend from the East.

"Something like that. Or the fall color in New England. All winter the experts take measurements and make forecasts. This year they predicted gold, but it's already gone platinum. In a spot where you'd expect a hundred flowers, we've got a thousand. More kinds than anybody alive has ever seen at once."

"But these are annual flowers?"

"Right."

"Well, then. . . ." Our nonbiologist friend struggled to frame her question: "If they weren't there *last* year, and this year they *are,* then who planted them?"

One of us blurted, "*God* planted them!"

We glanced at each other nervously: A picturesque response indeed, from scientifically trained types like ourselves. Yet it seemed more compelling than any pedestrian lecture on life cycles and latency periods. Where *had* they all come from? Had these seeds just been lying around in the dirt for decades? And how was it that, at the behest of some higher power than the calendar, all at once there came a crowd?

The answers to these questions tell a tale as complex as a Beethoven symphony. Before a concert, you could look at a lot of sheet music and try to prepare yourself mentally for the piece it inscribed, but you'd still be knocked out when you heard it performed. With wildflowers, as in a concert, the magic is in the timing, the subtle combinations—and, most important, the extent of the preparations.

For a species, the bloom is just the means to an end. The flower show is really about making seeds, and the object of the game is persistence through hell or high water, both of which are features of the Sonoran Desert. In winter, when snow is falling on much of North America, we get slow, drizzly rains that can last for days and soak the whole region to its core. The Navajo call these female rains, as opposed to the "male rains" of late summer—those rowdy thunderstorms that briefly disrupt the hot afternoons, drenching one small plot of ground while the next hill over remains parched. It's the female rains that affect spring flowering, and in some years, such as 1998, the benefaction trails steadily from winter on into spring. In others, after a lick and a promise, the weather dries up for good.

Challenging conditions for an ephemeral, these are. If a little seed begins to grow at the first promise of rain, and that promise gets broken, that right there is the end of its little life. If the same thing happened to every seed in the bank, it would mean the end of the species. But it *doesn't* happen that way. Desert wildflowers have had millennia in which to come to terms with their inconstant mother. Once the plant has rushed through growth and flowering, its seeds wait in the soil—and not just until the next time conditions permit germination, but often longer. In any given year, a subset of a species's seeds don't germinate, because they're programmed for a longer dormancy. This seed bank is the plant's protection against a beckoning rain followed by drought. If any kind of wildflower ever existed whose seeds all sprouted and died before following through to seed-set, then that species perished long ago. This is what natural selection is about. The species that have made it this far have encoded genetic smarts enough to outwit every peril. They produce seeds with different latency periods: Some germinate quickly, and some lie in wait, not just loitering there but loading the soil with many separate futures.

Scientists at the University of Arizona have spent years examining the intricacies of seed banks. Desert ephemerals, they've learned, use a surprising variety of strategies to fine-tune their own cycles to a climate whose cycles are not predictable—or at least, not predictable given the relatively short span of human observation. Even in a year as wet as 1998, when photo-ops and seed production exploded, the natives were not just seizing the moment; they were stashing away future seasons of success by varying, among and within species, their genetic schedules for germination, flowering, and seed-set. This variation reduces the intense competition that would result if every seed germinated at once. Some species even vary seed size: Larger seeds make more resilient sprouts, and smaller ones are less costly to produce; either morph may be programmed for delayed germination, depending on the particular strategy of the species. As a consequence of these sophisticated adaptations, desert natives can often hold their own against potential invasion by annual plants introduced from greener, more predictable pastures. You have to get up awfully early in the morning to outwit a native on its home turf.

The scientific term for these remarkable plants, "ephemeral annuals," suggests something that's as fragile as a poppy petal, a captive to the calendar. That is our misapprehension, along with our notion of this floral magic show—now you see it, now you don't—as a thing we can predict and possess like a garden. In spite of our determination to contain what we see in neat, annual packages, the blazing field of blues and golds is neither a beginning nor an end. It's just a blink, or maybe a smile, in the long life of a species whose blueprint for perseverance must outdistance all our record books. The flowers will go on mystifying us, answering to a clock that ticks so slowly we won't live long enough to hear it.

A Fist in the Eye of God

In the slender shoulders of the myrtle tree outside my kitchen window, a hummingbird built her nest. It was in April, the sexiest month, season of budburst and courtship displays, though I was at the sink washing breakfast dishes and missing the party, or so you might think. Then my eye caught a flicker of motion outside, and there

she was, hovering uncertainly. She held in the tip of her beak a wisp of wadded spiderweb so tiny I wasn't even sure it was there, until she carefully smoodged it onto the branch. She vanished then, but in less than a minute she was back with another tiny white tuft that she stuck on top of the first. For more than an hour she returned again and again, increasingly confident of her mission, building up by infinitesimal degrees a whitish lump on the branch—and leaving me plumb in awe of the supply of spiderwebbing on the face of the land.

I stayed at my post, washing everything I could find, while my friend did her own housework out there. When the lump had grown big enough—when some genetic trigger in her small brain said, "Now, that will do"—she stopped gathering and sat down on her little tuffet, waggling her wings and tiny rounded underbelly to shape the blob into a cup that would easily have fit inside my cupped hand. Then she hovered up to inspect it from this side and that, settled and waddled with greater fervor, hovered and appraised some more, and dashed off again. She began now to return with fine filaments of shredded bark, which she wove into the webbing along with some dry leaflets and a slap-dab or two of lichen pressed onto the outside for curb appeal. When she had made of all this a perfect, symmetrical cup, she did the most surprising thing of all: She sat on it, stretched herself forward, extended the unbelievable length of her tongue, and *licked* her new nest in a long upward stroke from bottom to rim. Then she rotated herself a minute degree, leaned forward, and licked again. I watched her go all the way around, licking the entire nest in a slow rotation that took ten minutes to complete and ended precisely back at her starting point. Passed down from hummingbird great-grandmothers immemorial, a spectacular genetic map in her mind had instructed her at every step, from snipping out with her beak the first spiderweb tuft to laying down whatever salivary secretion was needed to accrete and finalize her essential creation. Then, suddenly, that was that. Her busy urgency vanished, and she settled in for the long stillness of laying and incubation.

If you had been standing with me at my kitchen sink to witness all this, you would likely have breathed softly, as I did, "My God." The spectacular perfection of that nest, that tiny tongue, that beak calibrated perfectly to the length of the tubular red flowers from which she sucks nectar and takes away pollen to commit the essential act of copulation for the plant that feeds her—every piece of this thing and all of it, my God. You might be expressing your reverence for the details of a world created in seven days, 4,004 years ago (according to some biblical calculations), by a divine being approximately human in shape. Or you might be revering the details of a world created by a billion years of natural selection acting utterly without fail on every single life-form, one life at a time. For my money the latter is the greatest show on earth, and a church service to end all. I have never understood how anyone could have the slightest trouble blending religious awe with a full comprehension of the workings of life's creation.

Charles Darwin himself was a religious man, blessed with an extraordinary patience for observing nature's details, as well as the longevity and brilliance to put it all together. In his years of studying animate life he noticed four things, which any of us could notice today if we looked hard enough. They are:

1. Every organism produces more seeds or offspring than will actually survive to adulthood.
2. There is variation among these seeds or offspring.
3. Traits are passed down from one generation to the next.
4. In each generation the survivors succeed—that is, they survive—because they possess some advantage over the ones that don't succeed, and *because* they survive, they will pass that advantage on to the next generation. Over time, therefore, the incidence of that trait will increase in the population.

Bingo: the greatest, simplest, most elegant logical construct ever to dawn across our curiosity about the workings of natural life. It is inarguable, and it explains everything.

Most people have no idea that this, in total, is Darwin's theory of evolution. Furthermore, parents who tell their children not to listen to such talk because "it's just a theory" are ignorant of what that word means. A theory, in science, is a coherent set of principles used to explain and predict a class of phenomena. Thus, gravitational theory explains why objects fall when you drop them, even though it, too, is "just a theory." Darwin's has proven to be the most robust unifying explanation ever devised in biological science. It's stunning that he could have been so right—scientists of Darwin's time knew absolutely nothing about genetics—but he was. After a century and a half, during which time knowledge expanded boundlessly in genetics, geology, paleontology, and all areas of natural science, his simple logical construct continues to explain and predict perfectly the existence and behavior of every earthly life form we have ever studied. As the unifying principle of natural sciences, it is no more doubted among modern biologists than gravity is questioned by physicists. Nevertheless, in a bizarre recent trend, a number of states have limited or even outright banned the teaching of evolution in high schools, and many textbooks for the whole country, in turn, have wimped out on the subject. As a consequence, an entire generation of students is arriving in college unprepared to comprehend or pursue good science. Many science teachers I know are nostalgic for at least one aspect of the Cold War days, when *Sputnik* riveted us to the serious business of training our kids to real science, instead of allowing it to be diluted or tossed out to assuage the insecurities of certain ideologues.

We dilute and toss at our peril. Scientific illiteracy in our population is leaving too many of us unprepared to discuss or understand much of the damage we are wreaking on our atmosphere, our habitat, and even the food that enters our mouths. Friends of mine who opted in school for English lit instead of microbiology (an option I myself could easily have taken) sometimes come to me and ask, "In two hundred words or less, can you explain to me why I should be nervous about genetic engineering?" I tell them, "Sit down, I'll make you a cup of tea, and then get ready for more than two hundred words."

A sound-bite culture can't discuss science very well. Exactly what we're losing when we reduce biodiversity, the causes and consequences of global warming—these traumas can't be adequately summarized in an evening news wrap-up. Arguments *in favor* of genetically engineered food, in contrast, are dangerously simple: A magazine ad for an agribusiness touts its benevolent plan to "feed the world's hungry with our vitamin-engineered rice!" To which I could add in reply my own snappy motto: "If you thought that first free hit of heroin was a good idea. . . ." But before you can really decide whether or not you agree, you may need the five hundred words above and a few thousand more. If so, then sit down, have a cup of tea, and bear with me. This is important.

At the root of everything, Darwin said, is that wonder of wonders, genetic diversity. You're unlike your sister, a litter of pups is its own small Rainbow Coalition, and every grain of wheat in a field holds inside its germ a slightly separate destiny. You can't see the differences until you cast the seeds on the ground and grow them out, but sure enough, some will grow into taller plants and some shorter, some tougher, some sweeter. In a good year all or most of them will thrive and give you wheat. But in a bad year a spate of high winds may take down the tallest stalks and leave standing at harvest time only, say, the 10 percent of the crop that had a "shortness" gene. And if that wheat comprises your winter's supply of bread, plus the only seed you'll have for next year's crop, then you'll be almighty glad to have that small, short harvest. Genetic diversity, in domestic populations as well as wild ones, is nature's sole insurance policy. Environments change: Wet years are followed by droughts, lakes dry up, volcanoes rumble, ice ages dawn. It's a big, bad world out there for a little strand of DNA. But a population will persist over time if, deep within the scattered genetics of its ranks, it is literally prepared for anything. When the windy years persist for a decade, the wheat population will be overtaken by a preponderance of shortness, but if the crop maintains its diversity, there will always be recessive aspirations for height hiding in there somewhere, waiting to have their day.

How is the diversity maintained? That old black magic called sex. Every seed has two parents. Plants throw their sex to the wind, to a hummingbird's tongue, to the knees of a bee—in April you are *inhaling* sex, and sneezing—and in the process, each two parents put their scrambled genes into offspring that represent whole new genetic combinations never before seen on Earth. Every new outfit will be ready for *something,* and together—in a large enough population—the whole crowd will be ready for *anything.* Individuals will die, not at random but because of some fatal misfit between what an organism *has* and what's *required.* But the population will live on, moving always in the direction of fitness (however "fitness" is at the moment defined), not because anyone has a master plan but simply because survival carries fitness forward, and death doesn't.

People have railed at this reality, left and right, since the evening when a British ambassador's wife declared to her husband, "Oh dear, let us hope Mr. Darwin isn't right, and if he is, let us hope no one finds out about it!" Fundamentalist Christians seem disturbed by a scenario in which individual will is so irrelevant. They might be surprised to learn that Stalin tried to ban the study of genetics and evolution in Soviet universities for the opposite reason, attacking the idea of natural selection—which acts only at the level of the individual—for being anti-Communist. Through it all, the little engines of evolution have kept on turning as they have done for millennia, delivering us here and passing on, untouched by politics or what anybody thinks.

Nikolai Vavilov was an astounding man of science, and probably the greatest plant explorer who has ever lived. He spoke seven languages and could recite books by Pushkin from memory. In his travels through sixty-four countries between 1916 and 1940, he saw more crop diversity than anyone had known existed, and founded the world's largest seed collection.

As he combed continents looking for primitive crop varieties, Vavilov noticed a pattern: Genetic variation was not evenly distributed. In a small region of Ethiopia he found hundreds of kinds of ancient wheat known only to that place. A single New World plateau is astonishingly rich in corn varieties, while another one is rolling in different kinds of potatoes. Vavilov mapped the distribution of what he found and theorized that the degree of diversity of a crop indicated how long it had been grown in a given region, as farmers saved their seeds through hundreds and thousands of seasons. They also saved more *types* of seed for different benefits; thus popcorn, tortilla corn, roasting corn, and varieties of corn with particular colors and textures were all derived, over centuries, from one original strain. Within each crop type, the generations of selection would also yield a breadth of resistance to all types of pest and weather problems encountered through the years. By looking through his lens of genetics, Vavilov began to pinpoint the places in the world where human agriculture had originated. More modern genetic research has largely borne out his hypothesis that agriculture emerged independently in the places where the most diverse and ancient crop types, known as land races, are to be found: in the Near East, northern China, Mesoamerica, and Ethiopia.

The industrialized world depends entirely on crops and cultivation practices imported from what we now call the Third World (though evidently it was actually First). In an important departure from older traditions, the crops we now grow in the United States are extremely uniform genetically, due to the fact that our agriculture is controlled primarily by a few large agricultural corporations that sell relatively few varieties of seeds. Those who know the seed business are well aware that our shallow gene bank is highly vulnerable; when a crop strain succumbs all at once to a new disease, all across the country (as happened with our corn in 1970), researchers must return to the more diverse original strains for help. So we still rely on the gigantic insurance policy provided by the genetic variability in the land races, which continue to be hand-sown and harvested, year in and year out, by farmers in those mostly poor places from which our crops arose.

Unbelievably, we are now engaged in a serious effort to cancel that insurance policy.

It happens like this. Let's say you are an Ethiopian farmer growing a land race of wheat—a wildly variable, husky mongrel crop that has been in your family for hundreds of years. You always lose some to wind and weather, but the rest still comes through every year. Lately, though, you've been hearing about a kind of Magic Wheat that grows six times bigger than your crop, is easier to harvest, and contains vitamins that aren't found in ordinary wheat. And amazingly enough, by special arrangement with the government, it's free.

Readers who have even the slightest acquaintance with fairy tales will already know there is trouble ahead in this story. The Magic Wheat grows well the first year, but its rapid, overly green growth attracts a startling number of pests. You see insects on this crop that never ate wheat before, in the whole of your family's history. You watch, you worry. You realize that you're going to have to spray a pesticide to get this crop through to harvest. You're not so surprised to learn that by special arrangement with the government, the same company that gave you the seed for free can sell you the pesticide you need. It's a good pesticide, they use it all the time in America, but it costs money you don't have, so you'll have to borrow against next year's crop.

The second year, you will be visited by a terrible drought, and your crop will not survive to harvest at all; every stalk dies. Magic Wheat from America doesn't know beans about Ethiopian drought. The end.

Actually, if the drought arrived in year two and the end came that quickly, in this real-life fairy tale you'd be very lucky, because chances are good you'd still have some of your family-line seed around. It would be much more disastrous if the drought waited until the eighth or ninth year to wipe you out, for then you'd have no wheat left at all, Magic or otherwise. Seed banks, even if they're eleven thousand years old, can't survive for more than a few years on the shelf. If they aren't grown out as crops year after year, they die—or else get ground into flour and baked and eaten—and then this product of a thousand hands and careful selection is just gone, once and for all.

This is no joke. The infamous potato famine or Southern Corn Leaf Blight catastrophe could happen again any day now, in any place where people are once again foolish enough, or poor enough to be coerced (as was the case in Ireland), to plant an entire country in a single genetic strain of a food crop.

While agricultural companies have purchased, stored, and patented certain genetic materials from old crops, they cannot engineer a crop, *ever,* that will have the resilience of land races under a wide variety of conditions of moisture, predation, and temperature. Genetic engineering is the antithesis of variability because it removes the wild card—that beautiful thing called sex—from the equation.

This is our new magic bullet: We can move single genes around in a genome to render a specific trait that nature can't put there, such as ultrarapid growth or vitamin A in rice. Literally, we could put a wolf in sheep's clothing. But solving agricultural problems this way turns out to be far less broadly effective than the old-fashioned multigenic solutions derived through programs of selection and breeding. Crop predators evolve in quick and mysterious ways, while gene splicing tries one simple tack after another, approaching its goal the way Wile E. Coyote tries out each new gizmo from Acme only once, whereupon the roadrunner outwits it and Wile E. goes crestfallen back to the drawing board.

Wendell Berry, with his reliable wit, wrote that genetic manipulation in general and cloning in particular: ". . . besides being a new method of sheep-stealing, is only a pathetic attempt to make sheep predictable. But this is an affront to reality. As any shepherd would know, the scientist who thinks he has made sheep predictable has only made himself eligible to be outsmarted."

I've heard less knowledgeable people comfort themselves on the issue of genetic engineering by recalling that humans have been pushing genes around for centuries, through selective breeding of livestock and crops. I even read one howler of a quote that began, "Ever since Mendel spliced those first genes. . . ." These people aren't getting it, but I don't blame them—I blame the religious fanatics who kept basic biology out of their grade-school textbooks. Mendel did not *splice* genes, he didn't actually control anything at all; he simply watched peas to learn how their natural system of genetic recombination worked. The farmers who select their best sheep or grains to mother the next year's crop are working with the evolutionary force of selection, pushing it in the direction of their choosing. Anything produced in this way will still work within its natural evolutionary context of variability, predators, disease resistance, and so forth. But tampering with genes outside of the checks and balances you might call the rules of God's laboratory is an entirely different process. It's turning out to have unforeseen consequences, sometimes stunning ones.

To choose one example among many, genetic engineers have spliced a bacterium into a corn plant. It was arguably a good idea. The bacterium was *Bacillus thuringensis,* a germ that causes caterpillars' stomachs to explode. It doesn't harm humans, birds, or even ladybugs or bees, so it's one of the most useful pesticides we've ever discovered. Organic farmers have worked for years to expedite the path of the naturally occurring "Bt" spores from the soil, where the bacterium lives, onto their plants. You can buy this germ in a can at the nursery and shake it onto your tomato plants, where it makes caterpillars croak before sliding back into the soil it came from. Farmers have always used nature to their own ends, employing relatively slow methods circumscribed by the context of natural laws. But genetic engineering took a giant step and spliced part of the bacterium's DNA into a corn plant's DNA chain, so that as the corn grew, each of its cells would contain the bacterial function of caterpillar killing. When it produced pollen, each grain would have a secret weapon against the corn worms that like to crawl down the silks to ravage the crop. So far, so good.

But when the so-called Bt corn sheds its pollen and casts it to the wind, as corn has always done (it's pollinated by wind, not by bees), it dusts a fine layer of Bt pollen onto every tree and bush in the neighborhood of every farm that grows it—which is rapidly, for this popular crop, becoming the territory known as the United States. There it may explode the stomach of any butterfly larva in its path. The populations of monarch butterflies, those bold little pilgrims who migrate all the way to Mexico and back on wings the consistency of pastry crust, are plummeting fast. While there are many reasons for this (for example, their winter forests in Mexico are being burned), no reasonable person can argue that dusting them with a stomach explosive is going to help matters. So, too, go other butterflies more obscure, and more endangered. And if that doesn't happen to break your heart, just wait awhile, because something that pollinates your food and builds the soil underneath it may also be slated for extinction. And there's another practical problem: The massive exposure to Bt, now contained in every cell of this corn, is killing off all crop predators except those few that have mutated a resistance to this long-useful pesticide. As a result, those superresistant mutants are taking over, in exactly the same way that overexposure to antibiotics is facilitating the evolution of antibiotic-resistant diseases in humans.

In this context of phenomenal environmental upsets, with even larger ones just offstage awaiting their cue, it's a bit surprising that the objections to genetic engineering we hear most about are the human health effects. It is absolutely true that new combinations of DNA can create proteins we aren't prepared to swallow; notably, gene manipulations in corn unexpectedly created some antigens to which some humans are allergic. The potential human ills caused by ingestion of engineered foods remain an open category—which is scary enough in itself, and I don't mean to minimize it. But there are so many ways for gene manipulation to work from the inside to destroy our habitat and our food systems that the environmental challenges loom as something on the order of a cancer that might well make personal allergies look like a sneeze. If genetically reordered organisms escape into natural populations, they may rapidly change the genetics of an entire species in a way that could seal its doom. One such scenario is the "monster salmon" with genes for hugely rapid growth, which are currently poised for accidental release into open ocean. Another scenario, less cinematic but dangerously omnipresent, is the pollen escaping from crops, creating new weeds that we cannot hope to remove from the earth's face. Engineered genes don't play by the rules that have organized life for three billion years (or, if you prefer, 4,004). And in this case, winning means loser takes all.

Huge political question marks surround these issues: What will it mean for a handful of agribusinesses to control the world's ever-narrowing seed banks? What about the chemical dependencies they're creating for farmers in developing countries, where government deals with multinational corporations are inducing them to grow these engineered crops? What about the business of patenting and owning genes? Can there be any good in this for the flat-out concern of people trying to feed themselves? Does it seem *safe,* with the world now being what it is, to give up self-sustaining food systems in favor of dependency on the global marketplace? And finally, would *you* trust a guy in a suit who's never given away a nickel in his life, but who now tells you he's made you some *free* Magic Wheat? Most people know by now that corporations can do only what's best for their quarterly bottom line. And anyone who still believes governments ultimately do what's best for their people should be advised that the great crop geneticist Nikolai Vavilov died in a Soviet prison camp.

These are not questions to take lightly, as we stand here in the epicenter of corporate agribusiness and look around at the world asking, "Why on earth would they hate us?" The general ignorance of U.S. populations about who controls global agriculture reflects our trust in an assured food supply. Elsewhere, in places where people grow more food, watch less TV, and generally encounter a greater risk of hunger than we do, they mostly know what's going on. In India, farmers have persisted in burning to the ground trial crops of transgenic cotton, and they forced their government to ban Monsanto's "terminator technology," which causes plants to kill their own embryos so no viable seeds will survive for a farmer to replant in the next generation (meaning he'd have to buy new ones, of course). Much of the world has already refused to import genetically engineered foods or seeds from the United States. But because of the power and momentum of the World Trade Organization, fewer and fewer countries have the clout to resist the reconstruction of their food supply around the scariest New Deal ever.

Even standing apart from the moral and political questions—if a scientist *can* stand anywhere without stepping on the politics of what's about to be discovered—there are question marks enough in the science of the matter. There are consequences in it that no one knew how to anticipate. When the widely publicized Human Genome Project completed its mapping of human chromosomes, it offered an unsettling, not-so-widely-publicized conclusion: Instead of the 100,000 or more genes that had been expected, based on the number of proteins we must synthesize to be what we are, we have only about 30,000—about the same number as a mustard plant. This evidence undermined the central dogma of how genes work; that is, the assumption of a clear-cut chain of processes leading from a single gene to the appearance of the trait it controls. Instead, the mechanism of gene expression appears vastly more complicated than had been assumed since Watson and Crick discovered the structure of DNA in 1953. The expression of a gene may be altered by its context, such as the presence of other genes on the chromosome near it. Yet, genetic engineering operates on assumptions based on the simpler model. Thus, single transplanted genes often behave in startling ways in an engineered organism, often proving lethal to themselves, or, sometimes, neighboring organisms. In light of newer findings, geneticists increasingly concede that gene-tinkering is to some extent shooting in the dark. Barry Commoner, senior scientist at the Center for the Biology of Natural Systems at Queens College, laments that while the public's concerns are often derided by industry scientists as irrational and uneducated, the biotechnology industry is—ironically—conveniently ignoring the latest results in the field "which show that there are strong reasons to fear the potential consequences of transferring a DNA gene between species."

Recently I heard Joan Dye Gussow, who studies and writes about the energetics, economics, and irrationalities of global food production, discussing some of these problems in a radio interview. She mentioned the alarming fact that pollen from genetically engineered corn is so rapidly contaminating all other corn that we may soon have no naturally bred corn left in the United States. "This is a fist in the eye of God," she said, adding with a sad little laugh, "and I'm not even all that religious." Whatever you believe in—whether God for you is the watchmaker who put together the intricate workings of this world in seven days or seven hundred billion days—you'd be wise to believe the part about the fist.

Religion has no place in the science classroom, where it may abridge students' opportunities to learn the methods, discoveries, and explanatory hypotheses of science. Rather, its place is in the hearts of the men and women who study and then practice scientific exploration. Ethics can't influence the outcome of an experiment, but they can serve as a useful adjunct to the questions that get asked in the first place, and to the applications thereafter. (One must wonder what chair God occupied, if any, in the Manhattan Project.) In the halls of science there is often an unspoken sense that morals and objectivity can't occupy the same place. That is balderdash—they always have cohabited. Social norms and judgments regarding gender, race, the common good, cooperation, competition, material gain, and countless other issues reside in every active human mind, so they were hovering somewhere in the vicinity of any experiment ever conducted by a human. That is precisely why science invented the double-blind experiment, in which, for example, experimental subjects don't know whether they're taking the drug or the placebo, and neither does the scientist recording their responses, so as to avoid psychological bias in the results. But it's not possible to double-blind the scientist's approach to the task in the first place, or to the way results will be used. It is probably more scientifically constructive to acknowledge our larger agenda than to pretend it doesn't exist. Where genetic engineering is concerned, I would rather have ethics than profitability driving the program.

I was trained as a biologist, and I can appreciate the challenge and the technical mastery involved in isolating, understanding, and manipulating genes. I can think of fascinating things I'd like to do as a genetic engineer. But I only have to stand still for a minute and watch the outcome of thirty million years' worth of hummingbird evolution transubstantiated before my eyes into nest and egg to get knocked down to size. I have held in my hand the germ of a plant engineered to grow, yield its crop, and then murder its own embryos, and there I glimpsed the malevolence that can lie in the heart of a profiteering enterprise. There once was a time when Thoreau wrote, "I have great faith in a seed. Convince me that you have a seed there, and I am prepared to expect wonders." By the power vested in everything living, let us keep to that faith. I'm a scientist who thinks it wise to enter the doors of creation not with a lion tamer's whip and chair, but with the reverence humankind has traditionally summoned for entering places of worship: a temple, a mosque, or a cathedral. A sacred grove, as ancient as time.

Lily's Chickens

My daughter is in love. She's only five years old, but this is real. Her beau is shorter than she is, by a wide margin, and she couldn't care less. He has dark eyes, a loud voice, and a tendency to crow. He also has five girlfriends, but Lily doesn't care about that, either. She loves them all: Mr. Doodle, Jess, Bess, Mrs. Zebra, Pixie, and Kiwi. They're chickens. Lily likes to sit on an overturned bucket and sing to them in the afternoons. She has them eating out of her hand.

It began with coveting our neighbor's chickens. Lily would volunteer to collect the eggs, and then she offered to move in with them. Not the neighbors, the chickens. She said if she could have some of her own, she would be the happiest girl on earth. What parent could resist this bait? Our life style could accommodate a laying flock; my husband and I had kept poultry before, so we knew it was a project we could manage, and a responsibility Lily could handle largely by herself. I understood how much that meant to her when I heard her tell her grandmother, "They're going to be just *my* chickens, Grandma. Not even one of them will be my sister's." To be five years old and have some other life form entirely under your control—not counting goldfish or parents—is a majestic state of affairs.

So her dutiful father built a smart little coop right next to our large garden enclosure, and I called a teenage friend who might, I suspected, have some excess baggage in the chicken department. She raises championship show chickens, and she culls her flock tightly. At this time of year she'd be eyeing her young birds through their juvenile molt to be sure every feather conformed to the gospel according to the chicken-breeds handbook, which is titled, I swear, *The Standard of Perfection.* I asked if she had a few feather-challenged children that wanted adoption, and she happily obliged. She even had an adorable little bantam rooster that would have caused any respectable chicken-show judge to keel over—the love child of a Rose-comb and a Wyandotte. I didn't ask how it happened.

In Lily's eyes *this* guy, whom she named Mr. Doodle, was the standard of perfection. We collected him and a motley harem of sweet little hens in a crate and brought them home. They began to scratch around contentedly right away, and Lily could hardly bear to close her eyes at night on the pride she felt at poultry ownership. Every day after feeding them she would sit on her overturned bucket and chat with them about the important things.

She could do this for an hour, easily, while I worked nearby in the garden. We discovered that they loved to eat the weeds I pulled, and the grasshoppers I caught red-handed eating my peppers. We wondered, would they even eat the nasty green hornworms that are the bane of my tomato plants? *Darling*, replied Mrs. Zebra, licking her non-lips, *that was to die for.*

I soon became so invested in pleasing the hens, along with Lily, that I would let a fresh green pigweed grow an extra day or two to get some size on before pulling it. And now, instead of carefully dusting my tomato plants with Bacillus spores (a handy bacterium that gives caterpillars a fatal bellyache), I allow the hornworms to reach heroic sizes, just for the fun of throwing the chickens into conniptions. Growing hens alongside my vegetables, and hornworms and pigweeds as part of the plan, has drawn me more deeply into the organic cycle of my gardening that is its own fascinating reward.

Watching Mr. Doodle's emergent maturity has also given me, for the first time in my life, an appreciation for machismo. At first he didn't know what to do with all these girls; to him they were just competition for food. Whenever I tossed them a juicy bug, he would display the manners of a teenage boy on a first date at a hamburger joint, rushing to scarf down the whole thing, then looking up a little sheepishly to ask, "Oh, did you want some?" But as hormones nudged him toward his rooster imperatives, he began to strut with a new eye toward his coopmates. Now he rushes up to the caterpillar with a valiant air, picking it up in his beak and flogging it repeatedly against the ground until the clear and present danger of caterpillar attack has passed. Then he cocks his head and gently approaches Jess or Bess with a throaty little pickup line, dropping the defeated morsel at her feet. He doles out the food equitably, herds his dizzy-headed girls to the roost when it's time for bed, and uses an impressive vocabulary to address their specific needs: A low, monotonous cluck calls them to the grub; a higher-pitched chatter tells them a fierce terrestrial carnivore (our dog) is staring balefully through the chicken-wire pen; a quiet, descending croak warns "Heads up!" when the ominous shadow of an owl or hawk passes overhead. Or a dove, or a bumblebee—OK, this isn't rocket science. But he does his job. There is something very touching about Mr. Doodle when he stretches up onto his toes, shimmies his golden-feather shawl, throws back his little head, and cries—as Alexander Haig did in that brief moment when he thought he was president—"As of now, I *am* in control!"

With the coop built and chickens installed, all we had to do now was wait for our flock to pass through puberty and begin to give us our daily eggs. We were warned it might take a while because they would be upset by the move and would need time for emotional adjustment. I was skeptical about this putative pain and suffering; it is hard to put much stock in the emotional life of a creature with the I.Q. of an eggplant. Seems to me you put a chicken in a box, and she looks around and says, "Gee, life is a box." You take her out, she looks around and says, "Gee, it's sunny here." But sure enough, they took their time. Lily began each day with high hopes, marching out to the coop with cup of corn in one hand and my twenty-year-old wire egg-basket in the other. She insisted that her dad build five nest boxes in case they all suddenly got the urge at once. She fluffed up the straw in all five nests, nervous as a bride preparing her boudoir.

I was looking forward to the eggs, too. To anyone who has eaten an egg just a few hours' remove from the hen, those white ones in the store have the charisma of day-old bread. I looked forward to organizing my family's meals around the pleasures of quiches, Spanish tortillas, and soufflés, with a cupboard that never goes bare. We don't go to the grocery very often; our garden produces a good deal of what we eat, and in some seasons nearly all of it. This is not exactly a hobby. It's more along the lines of religion, something we believe in the way families believe in patriotism and loving thy neighbor as thyself. If our food ethic seems an unusual orthodoxy to set alongside those other two, it probably shouldn't. We consider them to be connected.

Globally speaking, I belong to the 20 percent of the world's population—and chances are you do, too—that uses 67 percent of the planet's resources and generates 75 percent of its pollution and waste. This doesn't make me proud. U.S. citizens by ourselves, comprising just 5 percent of the world's people, use a quarter of its fuels. An average American gobbles up the goods that would support thirty citizens of India. Much of the money we pay for our fuels goes to support regimes that treat their people—particularly their women—in ways that make me shudder. I'm a critic of this shameful contract, and of wasteful consumption, on general principles. Since it's nonsensical, plus embarrassing, to be an outspoken critic of things you do yourself, I set myself long ago to the task of consuming less. I never got to India, but in various stages of my free-wheeling youth I tried out living in a tent, in a commune, and in Europe, before eventually determining that I could only ever hope to dent the salacious appetites of my homeland and make us a more perfect union by living *inside* this amazing beast, poking at its belly from the inside with my one little life and the small, pointed sword of my pen. So this is where I feed my family and try to live lightly on the land.

The Union of Concerned Scientists notes that there are two main areas where U.S. citizens take a hoggish bite of the world's limited resources and fuels. First is transportation. Anybody would guess this. I'm lucky, since I can commute from bedroom to office in my fuzzy slippers, by way of the coffeepot in the kitchen. We get the kids to school via bus and carpool and organize our errands so trips to town are minimized. I have lived some years of my adulthood without a car (it's easier in Europe), though for now I have one. I hope soon to trade it in for one of those electric-hybrid station wagons that gets forty-eight miles per gallon. Ironically, my interests in conservation and the personal act as political have led me into a career that garners me hundreds of invitations a year to burn jet fuel in order to spread my gospel. I solve this dilemma, imperfectly, by sticking mostly to recycled paper as the medium of that gospel and turning down ninety-nine invitations out of a hundred, taking only the trips that somehow promise me a story whose telling will have been worth its purchase. So in the realm of transporting myself, so long as I can avoid the wild-goose chase of a book tour, I can live within fairly modest means.

Gas-guzzling area number two, and this may surprise you, is our diet. Americans have a taste for food that's been seeded, fertilized, harvested, processed, and packaged in grossly energy-expensive ways and then shipped, often refrigerated, for so many miles it might as well be green cheese from the moon. Even if you walk or bike to the store, if you come home with bananas from Ecuador, tomatoes from Holland, cheese from France, and artichokes from California, you have guzzled some serious gas. This extravagance that most of us take for granted is a stunning energy boondoggle: Transporting 5 calories' worth of strawberry from California to New York costs 435 calories of fossil fuel. The global grocery store may turn out to be the last great losing proposition of our species.

Most Americans are entangled in a car dependency not of our own making, but nobody *has* to eat foods out of season from Rio de Janeiro. It's a decision we remake daily, and an unnecessary kind of consumption that I decided some time ago to try to expunge from my life. I had a head start because I grew up among farmers and have found since then that you can't take the country out of the girl. Wherever I've lived, I've gardened, even when the only dirt I owned was a planter box on an apartment balcony. I've grown food through good times and bad, busy and slow, richer and poorer—especially poorer. When people protest that gardening is an expensive hobby, I suggest they go through their garden catalogs and throw out the ones that offer footwear and sundials. Seeds cost pennies apiece or less. For years I've grown much of what my family eats and tried to attend to the sources of the rest. As I began to understand the energy crime of food transportation, I tried to attend even harder, eliminating any foods grown on the dark side of the moon. I began asking after the processes that brought each item to my door: what people had worked where, for slave wages and with deadly pesticides; what places had been deforested; what species were being driven extinct for my cup of coffee or banana bread. It doesn't taste so good when you think about what died going into it.

Responsible eating is not so impossible as it seems. I was encouraged in my quest by *This Organic Life*, a compelling book by Joan Dye Gussow that tells how, and more important *why,* she aspired to and achieved vegetable self-sufficiency. She does it in her small backyard in upstate New York, challenging me to make better use of my luxuries of larger space and milder clime. Sure enough, she's right. In the year since I started counting, I've found I need never put a vegetable on my table that has traveled more than an hour or so from its home ground to ours.

I should explain that I do this in the *places* where I live, because I am not I, but we. My husband and I met in our late thirties; he had already grown deep roots in a farming community in southern Appalachia. I had roots of my own, plus a kid, in my little rancho outside Tucson, Arizona. So our marriage is a more conspicuous compromise than most: We all live out the school year in the Southwest and spend the summer growing season in Appalachia. By turns we work two very different farms, both of which we share with other families who inhabit them year-round so nothing has to lie very fallow or stand empty. Eventually, when we've fulfilled all our premarital obligations, we'll settle in one place. Until then I blow some of the parsimony of my daily bedroom-slipper commute on one whopper of an annual round trip, but it's a fine life for a gardener. In the mild winters of Tucson, where we get regular freezes but no snow, we grow the cool-weather crops that can take a little frost: broccoli, peas, spinach, lettuce, Chinese vegetables, garlic, artichokes. And in the verdant southern summers, we raise everything else: corn, peppers, green beans, tomatoes, eggplants, too much zucchini, and never enough of the staples (potatoes, dried beans) that carry us through the year. Most of whatever else I need comes from the local growers I meet at farmers' markets. Our family has arrived, as any sentient people would, at a strong preference for the breads and pasta we make ourselves, so I'm always searching out proximate sources of organic flour. Just by reading labels, I have discovered I can buy milk that comes from organic dairies only a few counties away; in season I can often get it from my neighbors, in exchange for vegetables; and I've become captivated by the alchemy of creating my own cheese and butter. (Butter is a sport; cheese is an art.) Winemaking remains well beyond my powers, but fortunately good wine is made in both Arizona and Virginia, and in the latter state I am especially glad to support some neighbors in a crashing tobacco-based economy who are trying to hold on to their farms by converting them to vineyards. Somewhere near you, I'm sure, is a farmer who desperately needs your support, for one of a thousand reasons that are pulling the wool out of the proud but unraveling traditions of family farming.

I am trying to learn about this complicated web as I go, and I'm in no position to judge anyone else's personal habits, believe me. My life is riddled with energy inconsistencies: We try hard to conserve, but I've found no way as yet to rear and support my family without a car, a computer, the occasional airplane flight, a teenager's bathroom equipped with a hair dryer, et cetera. I'm no Henry D. Thoreau. (And just for the record, for all his glorification of his bean patch, Henry is known habitually to have gone next door to eat Mrs. Ralph W. Emerson's cooking.) Occasional infusions of root beer are apparently necessary to my family's continued life, along with a brand of vegetable chips made in Uniondale, New York. And there's no use in my trying to fib about it, either, for it's always when I have just these items in the grocery cart, and my hair up in the wackiest of slapdash ponytails, that some kind person in the checkout line will declare, "Oh, Ms. Kingsolver, I just love your work!"

Our quest is only to be thoughtful and simplify our needs, step by step. In the way of imported goods, I try to stick to nonperishables that are less fuel-costly to ship; rice, flour, and coffee are good examples. Just as simply as I could buy coffee and spices from the grocery, I can order them through a collective in Fort Wayne, Indiana, that gives my money directly to cooperative farmers in Africa and Central America who are growing these crops without damaging their tropical habitat. We struggled with the notion of giving up coffee altogether until we learned from ornithologist friends who study migratory birds being lost to habitat destruction, that there is a coffee-cultivation practice that helps rather than hurts. Any coffee labeled "shade grown"—now available in most North American markets—was grown under rain-forest canopy on a farm that is holding a piece of jungle intact, providing subsistence for its human inhabitants and its birds.

I understand the power implicit in these choices. That I have such choices at all is a phenomenal privilege in a world where so many go hungry, even as our nation uses food as a political weapon, embargoing grain shipments to places such as Nicaragua and Iraq. I find both security and humility in feeding myself as best I can, and learning to live within the constraints of my climate and seasons. I like the challenge of organizing our meals as my grandmothers did, starting with the question of season and which cup is at the moment running over. I love to trade recipes with my gardening friends, and join in their cheerful competition to see who can come up with the most ways to conceal the i.d. of a zucchini squash.

If we are blessed with an abundance of choices about food, we are surely also obliged to consider the responsibility implicit in our choices. There has never been a more important time to think about where our food comes from. We could make for ourselves a safer nation, overnight, simply by giving more support to our local food economies and learning ways of eating and living around a table that reflects the calendar. Our families, of course, will never need to be as beholden to the seasons as the Native Americans who called February by the name "Hungry Month," and I'm grateful for that. But we can try to live close enough to the land's ordinary time that we notice when something is out of place and special. My grandfather Kingsolver used to tell me with a light in his eyes about the boxcar that came through Kentucky on the L&N line when he was a boy—only once a year, at Christmas—carrying oysters and oranges from the coast. Throughout my own childhood, every year at Christmastime while an endless burden of wants burgeoned around everybody else, my grandfather wanted only two things: a bowl of oyster soup and an orange. The depth of his pleasure in that meal was so tangible, even to a child, that my memory of it fills me with wonder at how deeply fulfillment can blossom from a cultivated ground of restraint.

I remember this as I struggle—along with most parents I know—to make clear distinctions between love and indulgence in raising my children. I honestly believe that material glut can rob a child of certain kinds of satisfaction—though deprivation is no picnic, either. And so our family indulges in exotic treats on big occasions. A box of Portuguese clementines one Christmas is still on Lily's catalog of favorite memories, and a wild turkey we got from Canada one Thanksgiving remains on my own. We enjoy these kinds of things spectacularly because at our house they're rare.

And yes, we eat some animals, in careful deference to the reasons for avoiding doing so. I don't really feel, as some have told me, that it's a sin to eat anything with a face, nor do I believe it's possible to live by that rule unless one maintains a certain degree of purposeful ignorance. Butterflies and bees and locusts all have faces, and they die like lambs to the slaughter (and in greater numbers) whenever a field of vegetable food is sprayed or harvested. Faceless? Not the birds that eat the poisoned insects, the bunnies sliced beneath the plow, the foxes displaced from the forest-turned-to-organic-wheat-field, and so on. If the argument is that meat comes from *higher orders* of life than those creatures, I wonder how the artificial, glassy-eyed construct of a bovine life gets to weigh more than the wiles of a fox or the virtuosity of a songbird. Myself, I love wild lives at least as much as tame ones, and eating costs lives. Even organic farmers kill crop predators in ways that aren't pretty, so a vegetable diet doesn't provide quite the sparkling karma one might wish. Most soybeans grown in this country are genetically engineered in ways that are anathema to biodiversity. So drinking soy milk, however wholesome it may be, doesn't save animals.

No, it's the other savings that compel me most toward a vegetable-based diet—the ones revealed by simple math. A pound of cow or hog flesh costs about ten pounds of plant matter to produce. So a field of grain that would feed a hundred people, when fed instead to cows or pigs that are *then* fed to people, fills the bellies of only ten of them; the other ninety, I guess, will just have to go hungry. That, in a nutshell, is how it's presently shaking down with the world, the world's arable land, and the world's hamburger eaters.

Some years ago our family took a trip across the Midwest to visit relatives in Iowa, and for thousands of miles along the way we saw virtually no animal life except feedlots full of cattle—surely the most unappetizing sight and smell I've encountered in my life (and my life includes some years of intimacy with diaper pails). And we saw almost no plant life but the endless fields of corn and soybeans required to feed those pathetic penned beasts. Our kids kept asking, mile after mile, "What used to be here?" It led to long discussions of America's vanished prairie, Mexico's vanished forests, and the diversity of species in the South American rain forests that are now being extinguished to make way for more cattle graze. We also talked about a vanishing American culture: During the last half century or so, each passing year has seen about half a million more people move away from farms (including all of my children's grandparents or great-grandparents). The lively web of farmhouses, schoolhouses, pasture lands, woodlots, livestock barns, poultry coops, and tilled fields that once constituted America's breadbasket has been replaced with a meat-fattening monoculture. When we got home our daughter announced firmly, "I'm never going to eat a cow again."

When your ten-year-old calls your conscience to order, you show up: She *hasn't* eaten a cow since, and neither have we. It's an industry I no longer want to get tangled up in, even at the level of the ninety-nine-cent exchange. Each and every quarter pound of hamburger is handed across the counter after the following productions costs, which I've searched out precisely: 100 gallons of water, 1.2 pounds of grain, a cup of gasoline, greenhouse-gas emissions equivalent to those produced by a six-mile drive in your average car, and the loss of 1.25 pounds of topsoil, every inch of which took five hundred years for the microbes and earthworms to build. How can all this cost less than a dollar, and who is supposed to pay for the rest of it? If I were a cow, right here is where I'd go mad.

Thus our family parted ways with all animal flesh wrought from feedlots. But for some farmers on certain land, assuming that they don't have the option of turning their acreage into a national park (and that people will keep wanting to eat), the most ecologically sound use of it is to let free-range animals turn its grass and weeds into edible flesh, rather than turning it every year under the plow. We also have neighbors who raise organic beef for their family on hardly more than the byproducts of other things they grow. It's quite possible to raise animals sustainably, and we support the grass-based farmers around us by purchasing their chickens and eggs.

Or we did, that is, until Lily got her chickens. The next time a roasted bird showed up on our table she grew wide-eyed, set down her fork, and asked, "Mama . . . is that . . . Mr. Doodle?"

I reassured her a dozen times that I would *never* cook Mr. Doodle; this was just some chicken *we didn't know*. But a lesson had come home to, well, roost. All of us sooner or later must learn to look our food in the face. If we're willing to eat an animal, it's probably only responsible to accept the truth of its living provenance rather than pretending it's a "product" from a frozen-foods shelf with its gizzard in a paper envelope. I've been straight with my kids ever since the first one leveled me with her eye and said, "Mom, no offense, but I think *you're* the Tooth Fairy." So at dinner that night we talked about the biology, ethics, and occasional heartbreaks of eating food. I told Lily that when I was a girl growing up among creatures I would someday have to eat, my mother had promised we would never butcher anything that had a first name. Thereafter I was always told from the outset which animals I could name. I offered Lily the same deal.

So she made her peace with the consumption of her beloveds' nameless relatives. We still weren't sure, though, how we'd fare when it came to eating their direct descendants. We'd allowed that next spring she might let a hen incubate and hatch out a few new chicks (Lily quickly decided on the precise number she wanted and, significantly, their names), but we stressed that we weren't in this business to raise ten thousand pets. Understood, said Lily. So we waited a week, then two, while Jess, Bess, and company worked through their putative emotional trauma and settled in to laying. We wondered, How will it go? When our darling five-year-old pantheist, who believes that even stuffed animals have souls, goes out there with the egg basket one day and comes back with eggs, how will we explain to her that she can't name those babes, because we're going to scramble them?

Here is how it went: She returned triumphantly that morning with one unbelievably small brown egg in her basket, planted her feet on the kitchen tile, and shouted at the top of her lungs, "Attention, everybody, I have an announcement: FREE BREAK-FAST."

We agreed that the first one was hers. I cooked it to her very exact specifications, and she ate it with gusto. We admired the deep red-orange color of the yolk, from the beta-carotenes in those tasty green weeds. Lily could hardly wait for the day when all of us would sit down to a free breakfast, courtesy of her friends. I wish that every child could feel so proud, and every family could share the grace of our table.

I think a lot about those thirty citizens of India who, it's said, could live on the average American's stuff. I wonder if I could build a life of contentment on their material lot, and then I look around my house and wonder what they'd make of mine. My closet would clothe more than half of them, and my books—good Lord—could open a library branch in New Delhi. Our family's musical instruments would outfit an entire (if very weird) village band, featuring electric guitars, violin, eclectic percussion section, and a really dusty clarinet. We have more stuff than we need; there is no question of our being perfect. I'm not even sure what "perfect" means in this discussion. I'm not trying to persuade my family to evaporate and live on air. We're here, we're alive, it's the only one we get, as far as I know, so I am keenly inclined to take hold of life by its *huevos*. As a dinner guest I gratefully eat just about anything that's set before me, because graciousness among friends is dearer to me than any other agenda. I'm not up for a guilt trip, just an adventure in bearable lightness. I approach our efforts at simplicity as a novice approaches her order, aspiring to a lifetime of deepening understanding, discipline, serenity, and joy. Likening voluntary simplicity to a religion is neither hyperbole nor sacrilege. Some people look around and declare the root of all evil to be sex or blasphemy, and so they aspire to be pious and chaste. Where I look for evil I'm more likely to see degradations of human and natural life, an immoral gap between rich and poor, a ravaged earth. At the root of these I see greed and overconsumption by the powerful minority. I was born to that caste, but I can aspire to waste not and want less.

I'm skeptical of evangelism, so I'm not going to have a tent revival here. But if you've come with me this far, you are in some sense a fellow traveler, and I'm glad for your company. In this congregation we don't confess or sit around chanting "we are not worthy"; we just do what we can and trust that the effort matters. Of all the ways we consume, food is a sensible one to attend to. Eating is a genuine need, continuous from our first day to our last, amounting over time to our most significant statement of what we are made of and what we have chosen to make of our connection to home ground. We can hardly choose *not* to eat, but we have to choose *how*, and our choices can have astounding consequences. Consider this: The average food item set before a U.S. consumer traveled 1,300 miles to get there. If Mr. Average eats ten or so items a day (and most of us eat more), in a year's time his food will have conquered five million miles by land, sea, and air. Picture a truck loaded with apples and oranges and iceberg lettuce rumbling to the moon and back ten times a year, all just for you. Multiply that by the number of Americans who like to eat—picture that flotilla of 285 million trucks on their way to the moon—and tell me you don't think it's time to revise this scenario.

Obviously, if you live in Manhattan, your child can't have chickens. But I'll wager you're within walking distance of a farmer's market where you can make the acquaintance of some farmers and buy what's in season. (I have friends in Manhattan who actually garden—on rooftops, and in neighborhood community plots.) In recent years nearly three thousand green markets have sprung up across the country, giving more than a hundred thousand farmers a place to sell their freshly harvested, usually organic produce to a regular customer base. In some seven hundred communities, both rural and urban (including inner-city New York), thousands of Americans are supporting their local food economies by signing up with Community-Supported Agriculture, a system that lets farmers get paid at planting time for produce that they then deliver weekly to their subscribers until year's end. Thousands of other communities have food co-operatives that specialize at least in organic goods, if not local ones, and promote commodities (such as bulk flours, cereals, oils, and spices) that minimize energy costs for packaging and shipping. Wherever you are, if you have a grocery store, you'll find something in there that is in season and hasn't spent half its life in a boxcar. The way to find out is to *ask*. If every U.S. consumer would earmark just ten dollars a month for local items, the consequences would be huge.

I realize there are deep, traditional divisions of class between white bread and whole wheat. I grew up among many people who would feel uncomfortable saying the word *organic* out loud. But I know I am witnessing a reordering of tradition when some of my rural Virginia neighbors who've heretofore grown, and chewed, tobacco become comfortable saying (and growing) "Chardonnay" and "Merlot grape." A dear friend of mine who has gardened for over six decades using the fertilizers and pesticides recommended by her farming father and husband, while they lived, confided to me not long ago that she'd secretly gone organic. (Her tomatoes that summer were some of her best ever.) It's clear that this movement is reaching across class lines, when farmers' markets redeem more than $100 million in food stamps each year. Community food-security initiatives in many areas are also working to link organic farmers with food banks and school lunch programs. Growing and eating are both infused with new politics.

Before anyone rules out eating locally and organically because it seems expensive, I'd ask him or her to figure in the costs paid *outside* the store: the health costs, the land costs, the big environmental Visa bill that sooner or later comes due. It's easy to notice that organic vegetables cost more than their chemically reared equivalents, but that difference is rarely the one consumers take home. A meal prepared at home from whole, chemical-free ingredients costs just pennies on the dollar paid for the highly processed agribusiness products that most Americans eat at restaurants or heat up in the microwave nearly every day. For every dollar we send to a farmer, fisherman, or rancher, we send between three and four to the shippers, processors, packagers, retailers, and advertisers. And there are countless other costs for that kind of food. Our history of overtaking the autonomy and economies of small countries with our large corporations, the wars and campaigns we wage to maintain our fossil-fuel dependency—these have finally brought us costs beyond our wildest fears. Cancer is expensive, too, as are topsoil loss and species extinction. The costs of global warming will bring us eventually to our knees. When I have to explain to my kids someday that, yes, back at the turn of the century we *did* know we were starting to cause catastrophic changes in the planet's climate that might end their lives prematurely, do I have to tell them we just couldn't be bothered to alter our convenience-food habits?

It doesn't, in principle, take more time to buy a local peach than a world-weary banana, and cooking from whole ingredients is not prohibitively time-consuming, either. As a working mother I am possessive of my time; I have to log in hours on my job—about forty a week—my spouse does the same, and our kids require of us the usual amount of kid-attention. But sometimes our family outings involve picking apples. I can peel the fruit and cook it into pies, jam, and purees for flavoring yogurt while I listen to the news on the radio or hear about my kids' day at school. Like many busy families, we cook in quantity on the weekends and freeze portions for easy midweek dinners. And we've befriended some fascinating microbes that will stay up all night in our kitchen making yogurt, feta, neufchatel, and sourdough bread without adult supervision. (I think copulation is involved, but we're open-minded.) Gardening is the best way I know to stay fit and trim, so during garden season, when it's up to me to make the earth move, I don't waste hours at the gym. Eating this way requires organization and skills more than time. Our great-grandmas did all this, and they may not have had other employment, but they did have to skin hogs for shoe leather, cut stove wood, sew everybody's clothes, and make the soap to wash them. Sheesh. My kitchen's on Easy Street.

It seems to me that giving up junk foods and jet-lagged vegetables is something like giving up smoking: It takes some discipline at first, but in the long run it's hard to see the minus sign in the equation. If there's anyone left who still thinks eating organically is a bland, granola-crunching affair, he or she must have missed the boat back around midmorning in the Age of Aquarius. The movement has grown up. Most Europeans think we're fools to eat some of the tasteless gunk that passes for food in our supermarkets. The Italians who pioneered Slow Food have forged a conscientious movement for preserving farms and the culture of unique, sustainable foods, but their starting point was pure epicurean disgust with fast food and watery, transported vegetables. Now that I've gotten into local eating I can't quit, because I've inadvertently raised children who are horrified by the taste of a store-bought tomato. Health is an issue, too: My growing girls don't need the hormones and toxins that lace American food in regulated quantities (the allowable doses are more about economic feasibility than about proven safety). But that is only part of the picture. Objecting to irresponsible agriculture for reasons of your personal health is a bit like objecting to having a nuclear power plant in your backyard for reasons of your view. My own two children are the smallest part of the iceberg. The millions of children in sub-Saharan Africa and other places now facing famine and historically unprecedented climatic extremes because of global warming—they are the rest of the iceberg.

Developing an intimate relationship with the processes that feed my family has brought me surprising personal rewards. I've tasted heirloom vegetables with poetic names—Mortgage Lifter tomato, Moon and Stars watermelon—whose flavors most never will know because they turn to pulp and vinegar in a boxcar. I've learned how to look a doe-goat right in the weird horizontal pupil of her big brown eye, sit down and extract her milk, and make feta cheese. (Step 1 is the hardest.) I've learned that with an unbreakable jar and the right music, a gang of kids can render butter from cream in eleven minutes flat. I've discovered a kind of citrus tree that withstands below-zero temperatures, almost extinct today but commonly grown by farm wives a hundred years ago. I've learned that the best-tasting vegetables on God's green earth are the ones our garden-wise foremothers bred for consumption, not hard travel. And I seem to be raising kids who like healthy food. When Lily streaks through the crowd at the farmer's market shouting, "Mama, look, they have *broccoli*, let's get a *lot!*"—well, heads do turn. Women have asked me, "How do you get one like that?"

I'm not going to tell you it's a done deal. If there were a bin of Twinkies at the farmer's market, the broccoli would go to rot.

Once upon a time, when I had my first baby, I believed that if I took care not to train her to the bad habits of sugar, salt, and fat, she would grow up not wanting those things. That delusion lasted exactly one year, until someone put a chocolate-frosted birthday confection in front of my sugar-free child and—how can I say this delicately?—she put her face in the cake. We humans crave sugar, fat, and salt because we evolved through thousands of years in which these dietary components were desperately scarce; those members of the tribe who most successfully glutted on them, when they found them, would store up the body fat to live through lean times and bear offspring. And now we've organized the whole enchilada around those latent biochemical passions—an early hominid's dream come true, a health-conscious mom's nightmare. If my cupboards were full of junk food, it would vanish, with no help from mice. We have our moments of abandon—Halloween, I've learned, is inescapable without a religious conversion—but most of the time my kids get other treats they've come to love. Few delicacies compare with a yellow-pear tomato, delicately sun-warmed and sugary, right off the vine. When I send the kids out to pick berries or fruit, I have to specify that at least *some* are supposed to go in the bucket. My younger daughter adores eating small, raw green beans straight off the garden trellis; I thought she was nuts till I tried them myself.

The soreness in my hamstrings at the end of a hard day of planting or hoeing feels good in a way that I can hardly explain—except to another gardener, who will know exactly the sweet ache I mean. My children seem to know it, too, and sleep best on those nights. I've found the deepest kind of physical satisfaction in giving my body's muscles, senses, and attentiveness over to the purpose for which they were originally designed: the industry of feeding that body and keeping it alive. I suspect that most human bodies have fallen into such remove from that original effort, we've precipitated an existential crisis that requires things like shopping, overeating, and adrenaline-rush movies to sate that particular body hunger.

And so I hope our family's efforts at self-provision will not just improve the health and habitat of my children but also offer a life that's good for them, and knowledge they need. I wish all children could be taught the basics of agriculture in school along with math and English literature, because it's surely as important a subject as these. Most adults my age couldn't pass a simple test on what foods are grown in their home counties and what month they come into maturity. In just two generations we've passed from a time when people almost never ate a fruit out of season to a near-universal ignorance of what seasons mean. One icy winter I visited a friend in Manhattan who described the sumptuous meal she was making for us, including fresh raspberries. "Raspberries won't grow in the tropics," I mused. "And they sure don't keep. So where would they come from in the dead of winter?" Without blinking she answered, "Zabar's!"

Apparently the guys running the show don't know much about agriculture, either, because the strategy of our nation is to run on a collision course with the possibility of being able to feed ourselves decently (or at all) in twenty years' time. I can't see how any animal could be this stupid; surely it's happening only because humans no longer believe food comes from dirt. Well, it does. Farmers are not just guys in overalls, part of the charming scenery of yesteryear; they are the technicians who know how to get teensy little seeds to turn into the stuff that comprises everything, and I mean *everything,* we eat. Is anybody paying attention? For every farm that's turned over to lawns and housing developments, a farmer is sent to work at the Nissan plant or the Kmart checkout line. What's lost with that career move is specific knowledge of how to gain food from a particular soil type, in a particular climate—wisdom that took generations to grow.

I want to protect my kids against a dangerous ignorance of

what sustains them. When they help me dig and hoe the garden, plant corn and beans, later on pick them, and later still preserve the harvest's end, compost our scraps, and then turn that compost back into the garden plot the following spring, they are learning important skills for living and maintaining life. I have also observed that they appreciate feeling useful. In fact, nearly all the kids I've ever worked with on gardening projects get passionate about putting seeds in the ground, to the point of earnest territoriality.

"Now," I ask them when we're finished, "what will you do if you see somebody over here tromping around or riding a bike over your seedbeds?"

"*We'll tell them to get outta our vegables!*" shouted my most recent batch of five-year-old recruits to this plot of mine for improving the world one *vegable* at a time.

Maria Montessori was one of the first child advocates to preach the wisdom of allowing children to help themselves and others, thereby learning to feel competent and self-assured. Most of the teachers and parents I know agree, and they organize classrooms and homes that promote this. But in modern times it's not easy to construct opportunities for kids to feel very useful. They can pick up their toys or take out the trash or walk the dog, but all of these things have an abstract utility. How useful is it to help take care of a dog whose main purpose, as far as they can see, is to be taken care of?

Growing food for the family's table is concretely useful. Nobody needs to explain how a potato helps the family. Bringing in a basket of eggs and announcing, "Attention, everybody: FREE BREAKFAST" is a taste of breadwinning that most kids can attain only in make-believe. I'm lucky I could help make my daughter's dream come true. My own wish is for world enough and time that every child might have this: the chance to count some chickens before they hatch.

The One-Eyed Monster, and Why I Don't Let Him In

"Nobody ever gets killed at our house," begins a song by Charlie King, and it continues with a litany of other horrors—"no one gets shot at, run over, or stabbed, / nobody goes up in flames"—that you'd surely agree you wouldn't want to see in *your* house, either, until you realize he's discussing what routinely happens on the screen that most people happily host in their living rooms. Maybe you have one in yours, and maybe you don't, but I'm with Charlie. People very rarely get killed at our house, and I'm trying to keep it that way.

The subject isn't entirely closed, of course, because we are not

Amish. We are what you'd call a regular American family, surrounded by regular America, and I believe in raising children who express themselves freely. This they do. The other night they raised the question once again of whether it might not be time for us to join the twenty-first century and every other upright-walking family we know of, at least in this neighborhood, and get cable TV.

"Why are you asking me?" I said, pretending to be dismayed. "Do I look like the dictator of this house?"

My efforts to stall weren't fooling anybody. I am not the dictator of this house, but I am the designated philosopher-king of its television-watching habits. That is to say, when my subjects become restless on the topic of TV, as they do from time to time, I sit down once again and explain to them in the kindest of tones why it is in their best interest to drop it.

But this time I'd been blindsided. Teenager and kindergartner were in league, with perhaps even the sympathies of my husband, though he was precluded from offering an opinion by his diplomatic ties. But the indentured serfs were fomenting a small rebellion.

"OK, look," I said to my serfs. "Watching TV takes *time*. When are you going to do it?"

They answered this without blinking: Evening. Morning. Prime time. Only when something good is on.

Which was just what I was afraid of. I explained that while I could understand there were probably some good things on TV that they were missing, they would have to miss out on *other* things in order to watch them, and when I looked around at what everybody was doing in our house, I couldn't really see what would give. I asked them, particularly my teenager (who likes to watch *Daria* and MTV at other people's houses, and whom I immediately sniffed out as our Robespierre here), to spend a few days paying careful attention to the hours of her life and exactly how she spent them. Kind of like keeping a calorie record, only with minutes. If she could come up with two expendable hours per day, I'd consider letting her spend them with the one-eyed monster.

She agreed to this, and at that moment I knew I'd already won. Here is what she does with her time: goes to school, does homework, practices the upright bass, talks with friends on the phone, eats dinner with the family, does more homework, reads for fun, hangs out with friends at their houses or ours, works out, listens to music, jams on the electric bass, tries to form an all-girl band, maintains various pets, participates in family outings, and gets exactly enough sleep. (In summertime the routine is different but the subject is moot, because then we live beyond the reach of cables, in a tiny house with no room for a TV and antique electricity that likely wouldn't support one anyway.) Her time card, in short, is full. Friends, exercise, music—not one minute of these would she give up, nor would I want her to. Even hanging out with friends—*especially* that—should not be sacrificed for solitary confinement with a talking box. If she wants to watch MTV at a friend's house, fine, that's *their* way of socializing—at our house her pals like to beat on my conga drums. And while she might have offered to trade in some hours of math homework, she knew better. Everything else she simply likes too much to cut out of her day.

So the discussion was shelved for the time being. I intend to keep a firm hand on at least this one aspect of my kingdom. To me, that ubiquitous cable looks an awful lot like the snake that batted its eyes at Eve.

Probably I shouldn't use such a morally loaded metaphor. I don't mean to equate my freedom from TV with freedom from sin, or to suggest that it confers on me any special virtue, though others generally interpret the discussion that way. If ever it comes up in conversation that my life is largely a TV void, people instantly get defensive about their own television-viewing habits and extol the value of the *few* things they like to watch (invariably citing something called the History Channel). But no defense is necessary, I promise; this is not about high-culture snobbery. If you knew me, you'd know there's almost nothing that is categorically beneath my dignity: I can get teary-eyed over a song about family reunions on the country radio station; I love to borrow my teenager's impractical shoes; on a dance floor I'm more at home with salsa and hip-hop than the tango; I have been known to do the Macarena. At a party more recent than I care to admit (I was definitely past forty), my friends voted me Most Likely to Dance on the Table. Before I dig this hole any deeper, why don't you just take my word for it? I'm not too high-minded for television, I really just don't like it. It's a taste I never got to acquire, having been raised by parents who made it painfully clear that life offered no bigger waste of time than watching the "boob tube" (one of the rare slang terms that've become more apt with the years). From there I proceeded to live an adult life with lots to do and very little cash, so that purposefully setting out to pay money for a time-wasting device just never crossed my mind. I made it to the childbearing phase without TV dependence, then looked around and thought, Well gee, why start *now?* Why get a pet python on the *day* you decide to raise fuzzy little gerbils?

The advantages of raising kids without commercial TV seem obvious, and yet I know plenty of parents who express dismay as their children demand sugar-frosted sugar for breakfast, then expensive name-brand clothing, then the right to dress up as hookers not for Halloween but for school. *Hello?* Anyone who feels powerless against the screaming voice of materialistic youth culture should remember that power comes out of those two little holes in the wall. The plug is detachable. Human young are not born with the knowledge that wearing somebody's name in huge letters on a T-shirt is a thrilling privilege for which they should pay eighty dollars. It takes years of careful instruction to arrive at that piece of logic.

I ask my kids, on principle, to live without wasteful and preposterous things (e.g., clothing that extorts from customers the right to wear labels on the outside), and it's a happier proposition to follow through if we don't have an extra blabbermouth in the room telling them every fourteen minutes about six brand-*new* wasteful, preposterous things they'll die without. It's fairly well documented that TV creates a net loss in contentment. The average household consumption of goods and services has doubled since 1957, when TV began to enter private homes, but according to a University of Chicago study, over all those years the fraction of Americans who describe themselves as "very happy" has remained steady, at about one third. The amount of money people believed they *needed* to buy happiness actually doubled between the mid-1980s and the mid-1990s, and the figure is still reaching for the sky. Think the kids would be unhappy without TV? I say pull the plug, quick, before they get more miserable. My daughters are by no means immune to peer pressure, but the kindergartner couldn't pick Tommy Hilfiger out of a lineup, and the ninth grader dresses way cool on an impressively frugal clothing budget, and they both find a hundred things to do each day that are more fun than sitting in front of a box. They agree with me about TV, once they're forced to accept the theory that there are only twenty-four hours in a day.

I can't think how anyone, child or adult, could sit still for the daily three hours and forty-six minutes that is our national TV-watching average. For my own purposes, sitting still is probably the most difficult part of the proposition. I have struggled all my life with a constitutional impatience with anything that threatens to waste what's left of my minutes here on earth. I start fidgeting at any community meeting where the first item on the agenda is to discuss and vote on the order of the other items on the agenda; I have to do discreet yoga relaxation postures in my chair to keep myself from hollering, "Yo, people, life is short!" I was born like this; I need to get a move on. I'm the kind of person who races around the kitchen so fast while cooking (with my mind on two other things), that I sometimes snag the fabric of my pants pocket on a drawer knob, and may either rip my trousers or fling the forks and spoons across the room. But reading can hold me spellbound, provided I'm the one turning the pages. And so I'm the kind of person who would rather *read* history (skipping over the parts I don't need) than have to sit and *watch* the eight-hour docudrama (baloney included). As a habitual reader I find the pace of information delivery on television noticeably sluggish. There's a perfectly good reason for this: The script for a one-hour TV documentary is only about fifteen or twenty double-spaced pages in length, whereas most any competent reader can cover three to five times that much material in an hour. Devoted as our culture is to efficiency, convenience, and DSL Internet access, I'm surprised so many Americans are content to get their up-to-the-minute news delivered in such a slow, vacuum-packed format.

I did get a chance, recently, to watch CNN for a few minutes, and I was bedazzled by what I presume to be its post–September 11 format, in which the main story is at the top of the screen, "Coming Up Next" occupies the middle, and completely unrelated headlines run constantly across the bottom. Yikes. It looked to me like a TV trying very hard to be a newspaper, about as successfully as my five-year-old imitates her big sister's smooth teenage dance steps. Ten minutes of that visual three-ring circus gave me a headache; when I want *newspaper,* I'll read one.

It's true that being a reader rather than a viewer gives me a type of naïveté that amuses my friends. I'm in the dark, for instance, about what many public figures look like, at least in color rather than newsprint (for the longest time I thought Phil Donahue was *blond*), and in some cases I may not know quite how to pronounce their names. When people began talking about the dreadful anthrax attack on the congressional offices, I kept scratching my head and asking, "*Dashell?* There's a Senator *Dashell?*" It took me nearly a day to identify the man whose name my brain had been registering as something like "Dask-lee."

But I don't mind being somebody's fool. I don't think I'm missing too much. Of course, every two weeks or so someone will tell me about the latest should-be-required-viewing-for-the-human-race documentary that I've missed. No problem: I know how to send off for it, usually from Annenburg CPB. Then some winter evening I'll put the tape in the machine, and if I agree it's wonderful I will see it out. Often it turns out I've long since read an article that told me exactly the same things about Muslim women or Mongolian mummies unearthed or whatever-have-you, or a book that told me more. For bringing events quickly to the world, the imaginative reporter's pad and the still photographer's snap are far more streamlined instruments than bulky video cameras and production committees. But sometimes, I won't deny it, there is a video image that stops me cold and rearranges the furniture of my heart. An African mother's gesture of resignation, the throat-singing of a Tuva shepherd, a silent pan of the untouched horizon of a Central Asian steppe—these things can carry an economy of feeling in so many unspoken words that they're pure instruction for a wordy novelist. For exactly this reason I love to watch movies, domestic and especially foreign ones, and we see lots of them at home. We do own a VCR. My kids would point out to you that it's old and somewhat antiquated; I would point out that so am I, by some standards, but we both still work just fine.

I'm happy to use the machine; I just don't want a cable or a dish or an antenna. Having a sieve up there on the roof collecting wild beams from everywhere does seem poetic, but the image that strikes me as more realistic is that of a faucet into the house that runs about 5 percent clear water and 95 percent raw sewage. I know some people who stay on guard all the time and carefully manage this flow so their household gets a healthy intake; I know a lot more who don't. Call me a control freak, but I have this thing about my household appliances—blender, lawn mower, TV monitor—which is that I like to feel I am in charge of the machine, and not vice versa. I have gotten so accustomed to this balance of power with my VCR that I behave embarrassingly in front of real television. When I watch those stony-faced men (I swear one of them is named Stone) deliver the official news with their pursed mouths and woeful countenances, I feel compelled to mutter back at them, insolently, while my teenager puts her forearms over her face. (Now you know what I really am—more insolent than my own teenager.) "You're completely ignoring what *caused* this," I mumble at Ted-Peter-Dan, "and anyway who did your *hair?*"

Well, honestly, who do those guys think I am? Thirteen seconds of whatever incident produced the most alarming visuals today, and I'm supposed to believe that's all I really need to know? One overturned fuel tanker in Nebraska is more important to me than, say, global warming? Television news is driven by compelling visuals, not by the intrinsic importance of the story being cast. Complicated, nonphotogenic issues requiring any considerable background information (global warming, for example) get left out of the running every time.

Meanwhile, viewers are lured into assuming, at least subconsciously, that this "news" is a random sampling of everything that happened on planet earth that day, and so represents reality. One friend of mine argued (even though, as I say, I'm not trying to start a fight) that he felt a moral obligation to watch CNN so he could see all there was and sort out what was actually true—as if CNN were some huge window thrown wide on the whole world at once. Not true, not *remotely* true. The world, a much wider place than seventeen inches, includes songbird migration, emphysema, pollinating insects, the Krebs cycle, my neighbor who recycles knitting-factory scraps to make quilts, natural selection, the Loess Hills of Iowa, and a trillion other things outside the notice of CNN. Are they important? Everything on that list I just tossed off is life or death to somebody somewhere, half of them are life and death to you and me, and you may well agree that they're all more interesting than Monica Lewinski. It's just a nasty, tiny subset of reality they're subsisting on there in TV land—the subset invested with some visual component likely to cause an adrenal reaction, ideally horror.

Print news has its multitude of agendas, to be sure, but they are not all so potently biased *against* the deeper assessment that interests me most. The overwhelming drive toward visuals in newscasting acts as a powerful influence on which bits of information will reach us. It also influences what we will retain. We are a predominantly visual species, and that's a biological fact that will never change; our brains are carefully wired to put the most stock in what they see, rather than what they hear. If we listen to a presidential candidates' debate over the radio, for instance, we'll be apt to recall, afterward, the visual components of the room in which we were sitting while we listened; if it was a fairly boring room, we'll also remember much of what the candidates said. If we watch that same debate on television, however, we will remember everything about the candidates' appearance—who was smug, who was tense, who made a funny face—but relatively few of their words. No matter what we may think of this prioritizing, it's the biological destiny of sighted people. It makes me wonder, frankly, why certain things are televised at all. If our aim is to elect candidates on the basis of their stature, clothing, and facial expressiveness, then fine, we should look at them. But if our intention is to evaluate their ideas, we should probably just listen and not look. Give us one good gander and we'll end up electing cheerleaders instead of careful thinkers. In a modern election, Franklin D. Roosevelt in his wheelchair wouldn't have a prayer—not to mention the homely but honest Abe Lincoln.

Still, there is this thing in us that wants to have a look, a curiosity that was quite useful to our ancestors on the savannah but is not so helpful now when it makes us rubberneck as we drive past the awful car crash. So the gods that gave us TV now bring us the awfulest car crash of the day and name it *The World Tonight.* This running real-time horror show provides a peculiarly unbalanced diet for the human psyche, tending to make us feel that we're living in the most dangerous time and place imaginable. When the eyes see a building explode, and then an airplane burning, and the ears hear, "Car bomb in Oklahoma City . . . far away from here . . . equipment failure . . . odds of this one in a million," the message stashed away by the brain goes something like, "Uh-oh, cars explode, buildings collapse, planes plunge to the ground—oh, *man*, better hunker down." As a person who either reads the news or hears it on the radio, I am a bit more of a stranger to this scary-world phenomenon, so I notice its impact on other people.

The day after all the world became a ghastly stage for the terrified high school students fleeing from murder by their classmates in Littleton, Colorado, it happened that I was to give an afternoon staff writing workshop at my sixth grader's school. When we assembled, I could see that the teachers were jumpy and wanted to talk rather than write. Several confessed that they had experienced physical panic that morning at the prospect of coming to school. I sympathized with their anxiety, but since nobody ever gets shot in my house, I didn't share the visceral sense of doom that surely came from seeing a live-camera feed of bloody children just like ours racing from a school so very much like this one. I remarked that while the TV coverage might make us *feel* endangered, the real probability of our own kids' getting shot at school today had been lower than the odds of their being bitten by a rattlesnake while waiting for the bus. And more to the point, the chance of such horror's happening here was hardly greater than it had been two days before, when we weren't remotely worried about it. (The TV coverage apparently did increase the likelihood of other school shootings, but only faintly.) It was such a small thing to offer—merely another angle on the truth—but I was amazed to see that it helped, as these thoughtful teachers breathed deeply, looked around at the quiet campus, and reclaimed the relative kindness of their lives. Anyone inclined toward chemical sedatives might first consider, seriously, turning off the TV. I know the vulnerability of my own psyche well enough to avoid certain films that are no doubt instructive and artful but will nevertheless insert violent images into my brain that I'll regret for many years. Obviously, I read verbal accounts of violence and construct from them my own mental pictures, but for whatever reason, these self-created images rarely have the same power as external ones to invade my mind and randomly, recurrently, savage my sense of wellbeing.

So I glean my news from many written sources and the radio, but even that isn't constant. I purposefully spend a few weeks each year avoiding national and international news altogether, and attending only to the news of my own community, since that is the only place I can actually do very much about the falling-apart-things of the moment. Some of my friends can't believe I do this, or can't understand it. One summer I was talking on the phone with a friend when she derived from our conversation that I had not yet heard about the tragic crash of the small airplane piloted by John Kennedy Jr.

"You're *kidding!*" she cried, again and again. "It happened three days ago, and you haven't heard about it yet?"

I hadn't.

My friend was amazed and amused. "People in Turkestan already know about this," she said.

I could have observed that everybody in the world, Turkestanis included, already knows global warming is the most important news on every possible agenda—except here in the United States, where that info has been successfully suppressed. We know so very much about the trees, and miss the forest. I was talking with a friend, though, so I told her only that I was deeply sorry for the Kennedy family, to whom this tragedy belonged, but that it would make no real difference in my life.

It's not that I'm callous about the calamities suffered by famous people; they are heartaches, to be sure, but heartaches genuinely experienced only by their own friends and families. It seems somewhat voyeuristic, and also absurd, to expect that JFK Jr.'s death should change my life any more than a recent death in *my* family affected the Kennedys. The same is true of a great deal—maybe most—of the other bad news that pounds at our doors day and night. On the matter of individual tragic deaths, I believe that those in my own neighborhood are the ones I need to attend to first, by means of casseroles and whatever else I can offer. I also believe it's possible to be so overtaken and stupefied by the tragedies of the world that we don't have any time or energy left for those closer to home, the hurts we should take as our own.

Many view this opinion as quaint. Truly, I'm in awe of the news junkies who can watch three screens at once and maintain their up-to-the-minute data without plunging into despair or cynicism. But I have a different sort of brain. For me, knowing does not replace doing. I find I sometimes need time off from the world of things I can't do anything about so I may be granted (as the famous prayer says) the serenity to accept the things I cannot change, the courage to change the things I can, and the wisdom to know the difference.

So for the duration of every summer, when our family migrates to a farm in rural Virginia (the place whose antique wiring would short out at the very idea of TV), I gather books and read up, seeking background information on the likes of genetic engineering, biodiversity, the history of U.S. relations in the Middle East—drifts of event too large and slow to be called news. I still listen almost daily to radio news (the Kennedy crash must not have been among All the Things to be Considered), but I limit myself to one national newspaper per week, usually the Sunday edition of the *Washington Post,* which I can buy on Monday or Tuesday at our little town's bookstore. Here's a big secret I've discovered that I will share with you now: This strategy saves me the time of reading about the sports hero/politician/movie star whose shocking assault charge/affair/heart attack was huge breaking news in the middle of last week, because by Sunday he has already confessed/apologized/died. You'd be amazed how little time it takes to catch up, not on "all the news that's fit to print," as one news organ boasts, but on all I really needed to be a responsible citizen.

The rest of the week I try to remember to stop by the hardware store and pick up our folkloric *County News,* which is comprised largely of farm forecasts and obituaries. I read them and make casseroles; it's a healthy exercise. It helps me remember what death really is, and helps me feel less useless in the face of it. And I decide with fresh conviction that we just don't need anybody getting killed at our house.

Letter to a Daughter at Thirteen

Here's a secret you should know about mothers: We spy. Yes, on our kids. It starts at birth. In those first months we spend twenty-three hours a day trying to get you to sleep, grateful you aren't yet verbal because at some point we run out of lyrics to the lullabies and start singing "Hush little baby, don't be contrary, / Mama's gonna have a coro-nary." And then you finally doze off, and what do you think we do? Go read a book? No, we stand over your cradle and stare, thinking, God, those little fingernails. Those eyelashes. Where did this perfect creature come from?

As you grow older, we attain higher orders of sneakiness. You're playing dolls with your friend, and we just pause outside the door of your room, *hmm-mm*, pretending to fiddle with the thermostat but really listening to you say, "Oh, my dear, here is your tea," as you hand her a recycled plastic Valvoline cap of pretend tea, and our hearts crack, we are such fools for love. We love you like an alcoholic loves gin—it makes our teeth hurt, it's the first thing we think about before we open our eyes in the morning—and like that, we take little swigs when nobody's looking.

These days I watch you while you're sitting at the table concentrating on algebra, running your hand through the blond curtain of your hair. Or after I've dropped you off at school and you've caught up to your friends, laughing, talking with your hands while your shoulders and hips rest totally at ease in the clothes and style you've made your own. I stare, wondering, How did I wind up with this totally cool person for a daughter?

You have confidence and wisdom beyond anything I'd found at your age. I thought of myself, at thirteen, as a collection of all the wrong things: too tall and shy to be interesting to boys. Too bookish. I had close friends, but I believed if I were a better person I would have more. At exactly your age I wrote in my diary, "Starting tomorrow I'm really going to try to be a better person. I have to change. I hope somebody notices." My diaries, whose first pages threatened dire punishment for anyone who snooped into them, would actually have slain any trespasser with pure boredom: I resolved with stupefying regularity to be good enough, better loved, happier. I looked high and low for the causes of my failure. I wrote poems and songs, then tore them up after unfavorable comparison with the work of Robert Frost or Paul Simon. My journal entries were full of a weirdly cheerful brand of self-loathing. "Dumb me" was how I christened any failure, regardless of its source. In a few years the perkiness would wane as I began to exhibit a genuine depression, beginning each day with desperate complaints about how hard it was to wake up, how I longed for nothing but sleep. I despaired of my ability to be liked by others or to accomplish anything significant, and I was stunned whenever anyone took any special interest in me.

Turning page after page in those old cardboard-bound diaries now, reading the faint penciled entries (I lacked even the confidence to use a pen), I dimly grasp in my memory the bleakness of that time. I feel such sadness now for that girl. This superachiever who started high school by winning a state essay contest and finished as valedictorian—why on earth did she fill her diary with the word *stupid*? What could any adult have said that would have helped? When I look at my yearbook photos, I'm surprised to see that I was pretty, for I certainly had no sense of it then. I put on the agreeable show I thought was required of a good girl, but I felt less valuable than everyone around me. I took small setbacks very hard. Every time I took a test, I predicted in my diary that I'd flunked it. I was like the anorexic girls who stare at their bony selves in a mirror and chant "I'm fat," except the ugliness was my very self. I chanted "Worthless me" while facing daily evidence to the contrary. I've always considered this to be the standard currency of adolescence. So it takes me by surprise when we're discussing some hassle and I sigh and say, "Adolescence is a pain," and you grin and reply, "Actually, it's not that bad."

As your maturity dawns over our relationship, I think hour by hour about how I was mothered and how I do the job myself. It doesn't explain the differences between my thirteen-year-old self and yours; I take no credit for your triumphs, nor was it my mother's fault that I was depressed. She did her best with a daughter who was surely frustrating. I remember her arguing with me, insisting almost angrily that I was pretty and talented and refused to see it. She must have rained steady compliments over my scholastic and artistic efforts. But compliments help only if one believes them. At some point before age thirteen, many girls stop believing in all praise, even when it comes straight from a mirror.

For you it's different. I watch you talking with your friends, or combing your little sister's hair, or standing at the back of your orchestra and elegantly bowing the strong bass line that holds everything else in place, and I see a quiet pride that's just part of your complexion. When you were little I used to declare you beautiful, and you'd smile and say, "I know." Now you're too savvy for that. But in the kitchen after school when you've reported something tough you dealt with well, and I say to you, "You have such good judgment about stuff like that," you'll look off to the side, and it'll be written all over your face: "I know." It's your prize possession. I'd do anything to see you keep it.

When I was pregnant with you, I read every book I could find on how to handle all things from diaper rash to warning lectures on sexually transmitted diseases. I became so appalled by the size of the task that I put my hands on my belly and thought, Oh Lord, can we just back up? But the minute you were born I looked at your hungry, squinched little face and *got* it: We do this thing one minute at a time. We'll never have to handle diaper rash and the sex lecture in the same day. My most important work will change from year to year, and I'll have time to figure it out. At first I was just Milk Central, then tiptoe walking coach and tea-party referee. Eventually I began to see that the common denominator, especially as mother of a girl child, was to protect and value every part of your personality and will, even when it differed from mine.

In this department I don't think girls of my generation got such a good shake from the guardians of our adolescence. The guidebook for parents then was organized around a whole different thesis; spanking was mandatory, and the word *self-esteem* had not been invented. The supervisors of my youth loved my accomplishments until I started campaigning against things they believed in. They thought I was beautiful, but they bluntly disparaged the getup required for *my* idea of beautiful. I wasn't even allowed to say I disliked a particular food. I made almost no significant decisions about my own life: I ate what I was fed, washed dishes but never planned meals, participated in school-sanctioned activities but virtually never hung out unsupervised with my friends. The parents of my time and place worried about pregnancy, drinking, and car accidents—as well they should have, since these shadows would fall sooner or later across the lives of most of my peers. I participated in a mind-boggling number of school-sanctioned activities but lacked time to be *me*, away from adults, just with peers. That must have looked too dangerous. As a child I'd spent endless hours poking around in the woods or playing disorganized games with other kids in the fields around our house, but once I grew breasts, my unchaperoned days were over. I felt increasingly scrutinized and failed to develop a natural ease or confidence with my peers. I was convinced that my parents would never let me grow up, so I railed against them internally but then felt guilty after, fearing they would mind-read my rebellious thoughts.

At age fifteen I was allowed to go on a trip with the high school English classes to see a performance of *Measure for Measure* in a nearby city. It was my first experience of Shakespeare (my first real play at all), and I felt elated afterward by this exposure to mature ideas and drama. But discussing it with my parents that night at dinner, I grew tense. There had been some implied sexuality in the play; my brother and I had made a pact not to mention it, but I feared somehow they knew anyway, and I was too nervous to eat. I felt sick inside, as if by watching this wonderful work, and loving it so much I'd betrayed my parents' trust in me and my own goodness.

When I went off to college at eighteen, I promptly went straight off the deep end of the social/recreational pool. It frightens me to look back on that reckless period of my life, but I also understand it perfectly. I'd been well under control up to that point, but I had no practice in *self*-control. I was extremely lucky not to damage myself in the process of learning moderation.

As penance for this close shave, I vowed early on to give you more choices than I had, so you could learn self-control in a safer laboratory than I did. The dance of letting go the reins is never easy—two steps forward, one step back. I've spent so much of my life stitching together the answers to the hard questions that it's natural for me to want to hand them down like a glove, one that will fit neatly onto an outstretched little clone hand. I try sometimes. But that glove won't fit. The world has changed, and even if it hasn't (drinking, drugs, and pregnancy are still at the top of the immediate-worry agenda), the answers will work for you only when you've stitched them together yourself.

People say it's because parents *love* their kids so much that they want to tell them how to live. But I'm afraid that's only half love, and the other half selfishness. Kids who turn out like their parents kind of validate their world. That was my first real lesson as a mother—realizing that you could be different from me, and it wouldn't make me less of a person. When you were three, in spite of all the toy socket wrenches and trucks I'd provided in my program of teaching you that women can be as capable and handy as men, you basically wanted to be the Princess Fairy Bride. You'd have given every one of your baby teeth for a Barbie doll. I tried to explain how this doll was an awful role model, she didn't look the way healthy women should, she was obsessed with clothes, blah blah. Translation: My worldview doesn't have room for Barbie in it, and I'd be embarrassed to have her as a houseguest. I wouldn't give in on Barbie.

Then one day you and your friend Kate were playing in your room, and I was spying just outside the door (yep, fiddling with the thermostat again) when I heard you say, "My mom won't let me have Barbies. But you know what? When I grow up I'm going to have *all the Barbie dolls I want!*"

Yikes, I thought to myself. Soon afterward, Barbie joined our family.

That was a stunner for me. Believe it or not, it was the first time I really pictured you as a someday-grown-up, completely in charge of yourself (and your menagerie of dolls). Eventually I'd have zero power over you, I realized, so this might be a good time to start preparing for it by shifting from 100 percent to 99 percent control. Let the Barbies come, and let you handle the Social Impact. You did, and along the way you probably learned a thing or two about physics: What happens when you shoot Barbie from a paper-towel tube? Also about disabilities: When the puppy found your abandoned Barbie party and left it looking like the Plane-Crash Barbie Close-Out Sale, I made you keep most of the Barbies, asking, "If your friend lost a leg or a hand, would you throw her away?" (The headless ones we laid to rest.) And I learned to say, when you dressed yourself in bridal veil, roller skates, rouge, and a tutu, "Wow, you have a really creative sense of style." I've never lied to you. I didn't say I thought you looked *good,* just creative. Maybe that's why you believe in my compliments now.

Every mom has to set limits, but that's never been so difficult with you. When you want something that I truly think will do you harm, I explain my reasons, and then usually let you have a *little* of it (except if it's illegal, or skydiving) or give you permission to abide by your friends' mothers' rules when you're at their houses (case in point: watching TV). Though you may not notice it, I'm keeping an eye out to see how long it takes you to decide you've had enough. Except for that one time when you put your whole face in the birthday cake, your judgment has proven exceptional.

All your life you've been apprenticing for adulthood. I recognized that when you were in preschool, learning how to be social: having feuds with girlfriends, then forgiving or sometimes moving on. One week they'd shun you, the next week you were queen bee while somebody else suffered. It tore me to pieces to watch, but I knew I couldn't save you. You were saving yourself, slowly. In fifth grade, it suddenly got harder: A boy started picking on you, mostly trying to embarrass you with sexual innuendo. Oh, man, did I want to walk into that classroom and knock some heads together. But I took a deep breath, knowing that even this—*especially* this—you had to learn to do for yourself. I was scared. It was my hardest mom event so far, and I didn't want to screw it up.

And it is *so* easy to screw this one up. When I was a teenager, the story I got from the world around me on how to behave with boys was a real song and dance, which boiled down to this: Boys want only one thing, which is to have sex with you, which is too nasty even to talk about, and it's your job to prevent it. They're also stronger than you and likely can do what they want, but if they succeed in raping you it's your fault, actually, because it was your job to avoid getting yourself into a position where you couldn't stop it. Also males are more important, they run the world, and if you want any kind of happiness or power, you're going to have to win their favor. Got it? Ready, set, go.

The day I sat down with you on your bed to talk about the Grade 5 boy problem, I felt as if I were jumping out of a speeding car, blindfolded, into a snake pit. I took a breath and said, "This is a good time for you to start learning how to handle inappropriate male attention." I told you three things: First, if you ever got truly scared, I would intervene. Second, it was fine to get really pissed off at this boy, because everybody deserved the right to go about her business without being harassed; the creepy feeling you had was *not* your fault, it was his. Third, boys are just people like us, and if they behave sensibly they can be very cool to be around—even in a physical way if that is your inclination, when you eventually feel the confidence and fondness to be with a guy like that.

Finally, I told you that unfortunately there would always be some guys who feel it's their gift to behave as irritants and scoundrels. You'd run into this many times in your life, and a classroom was a safer place to learn to defend yourself than, say, a college bar or a workplace.

Then we practiced role-playing. I wanted you to say, "No, I hate that, you make me sick, go away." You found it hard; your tendency was to be polite, even coy. I realized, with agony, that the world had already begun teaching you that girls should be pleased with, or at least politely tolerant of, male attention of any type. I tried not to hyperventilate. We practiced some more, you learned to take a very firm tone, and you made it through fifth grade. *I* learned what you were up against. It was not too early for me to begin thinking of you, and talking with you, as a transitional woman, with important disputed ground to claim for yourself on the map of equality. You've kept me posted on the main events in the boy-girl arena, and so far I've been impressed with how you've handled them.

I didn't do nearly so well myself, as a teenager. My first kiss happened the summer after I turned fourteen, at band camp—a school-sanctioned activity during which I was theoretically chaperoned every minute of the day. I met a cute boy named Dave who showed a flattering interest in me, and one evening when we were meant to be washing dishes he asked me to go outside instead, and mess around behind the so-called mess hall. I was scared to death; I went. Our kissing was nowhere near as graceful as the movies, with an icky dampness factor that seemed categorically not too different from washing dishes, but I felt thrilled to have been chosen. After camp ended I never heard from him again because, of course, we'd had no friendship, and I felt creepy about my tryst. I'm lucky he didn't expect me to go beyond kissing. I hope I'd have resisted (I'm pretty sure terror would have helped me out), but I'm sad to admit I can't say for certain. It took me years to get over being flattered and flattened by any kind of male approval. My first relationships in high school and early college were stunted by my inability to separate my interests from my boyfriend's. The guys who did time in that capacity during those years were invariably sweet; it wasn't as if they *meant* to ignore or malign me. It was just that I felt such pressure to remain coupled that I swallowed my own will to keep from rocking the boat. *Like* what he *likes*, *do* what he *wants:* I couldn't imagine just acting like myself in the company of a guy.

I see a lot of girls your age who are just the way I was then. I remember hearing one of your friends declare helplessly, "I can't say no to boys"—in the sixth grade! I feared for her future reproductive life. But not yours. I can see very well that if a male friend didn't take an interest in the things you care about, or wasn't respectful, you would use your remarkable charm and wit to lose him, fast. Or at least tell him that, as I heard you recently say, "he's not *all that* and a bag of potato chips." It's a huge relief to me. I look forward to meeting the guys you'll date.

You already know a lot of the things I had to teach myself in my late teens and early twenties. What saved me was nothing short of a complete transformation, the kind of soul-shattering revelation that some people find in religious salvation. *I* found it in the novels of Doris Lessing, Maxine Hong Kingston, Margaret Drabble, and Marilyn French, along with the words of Betty Friedan, Germaine Greer, Gloria Steinem, Robin Morgan, and lots of others. I began to find these books my last year in high school and then really sank into them in college, reading the way a drowning person breathes air when she finally breaks the surface. I stayed up late reading; I sat all day in the library on Saturdays, reading. Every word made sense to me, every claim brought me closer to being a friend to myself. These writers put names to the kinds of pain I'd been feeling for so long, the ways I felt useless in a culture in which women could be stewardesses but the pilots were all men. They helped me understand why I'd been so driven by the opinions of men. I was not stupid; in pandering to male favor I'd been pursuing what would be the smartest possible route to power in, say, Jane Austen's day, when women couldn't own property or vote. But these writers allowed me to imagine other possibilities. There are still many countries where women have to go the Jane Austen route: Muslim extremists stone women to death if they show their faces and declare their opinions in public, but here you'll only get some hate mail for it. The worst that was likely to happen to me, if I began standing up for myself at age nineteen, was that some guys who handled me with less deep concern than their auto transmissions would probably cut bait and run. This loss could be endured; that was all I needed to know. When my despair finally crystallized as anger, my conversion was rapid and absolute: I cut off my long hair, I began to dress for function rather than sexiness, I got mad at whosoever tried to bully me by virtue of unearned privilege—and I discovered there were guys who actually *liked* me this way. I joined a women's group on campus, then found a church that was more forgiving of personal lapses of judgment than of larger, social ones, such as war and hunger. I began working with migrant farm workers in central Indiana whose problems were larger than mine: They had no clean water or shelter. I learned more about the Vietnam war than I'd previously gleaned from *Reader's Digest*. By concentrating on what I could do to make things better for people who were worse off than me, I taught myself to feel significant. Word by word, day by day, I revised the word *stupid* out of my journal.

The premises of feminism—that women are entitled to do any kind of work men do, for the same pay, and to be accorded an equal measure of social respect—must seem obvious to you. But in 1973 these items were just barely on the agenda. The first time I suggested to my father that a woman could be president, he got a pained expression on his face just thinking about a woman having to go through that mess. He asked me, as delicately as he could, to consider what a disaster it would be if we had a war, and the president was on her menses. Both of us were acutely embarrassed, and that was the end of that. (It didn't occur to me until years later that most presidents are elected well past the age when menses would be an issue.) When I told my parents about an older college friend I admired who intended to keep her name after she got married, my mother offered sadly, "Any woman who'd do that doesn't love her husband."

My parents, in telling me of these and a thousand other limitations on my gender, weren't trying to hold me in contempt. They were merely advising me of the ways of the world—which, in 1973, held me in contempt. Since then they've changed their minds about many things, including my keeping my last name, which is now also yours. (And if you ever run for president, I'm positive Dad will vote for you.) But the persistence of misogyny in the world outside our family is not forgivable, and it makes me crazy. Why is it, for instance, that on the popular teen radio station, all the women are singing about guys who treat them like dirt (or, on a more optimistic note, declaring the jerk must go), while the men are chanting, sometimes literally, "Die, bitch, die!" It scares me that boys listen to this stuff; it scares me more that *girls* do. I can't tell you what to listen to, I know. To this day I get a buzz when I hear the first notes of "Lay Down Sally," probably from all the warnings I received against its morals and grammar. But if you're going to listen to these guys, *listen*. Eric Clapton was singing to me, "You are *so* the best, I can't stand for you to leave the room." I'd just once like to hear that from some rapper. One of the best gifts you ever gave me was when you turned off Eminem and started listening more to Sheryl Crow and Alanis Morisette on your CD player. I think—I hope—you did it not for me but for you. Because you didn't need "Die, bitch," as bedtime music.

I know that some girls of your acquaintance worship Eminem. Some are already doing drugs and having sex with guys because they need male approval that badly. I understand that perfectly, because of how *I* was in my teens. I wish I could tell them it's not too late yet: If they can just yank it back for a minute and find some little island of pride, there's hope. But it takes believing in some larger space for women in the world than they can presently see. For me, that belief came from the right books, because I happened to revere the printed word. Even more, it was finding and joining a huge, heady current that allowed me to believe I could change things a little—that I could fight back against what made me angry, in some way that was real and grown-up. Piercing and branding one's flesh or getting pregnant or getting AIDS is *not* fighting back, even though it may feel like it from the inside. From the outsider's point of view, these things make a display of self-loathing, which is the opposite of fighting back—it's a score for the opposition. I know, because I used to hate myself, and now I don't.

You never did, it seems. You like who you are, you work hard at whatever you do, you're kind to your friends, you show compassion for the world. You're a person I'd choose as a friend even if we weren't related. I actually like the ways you're turning out different from me; your confidence and smart-aleck wit inspire me. I was impressed, the day we were listening to the presidential campaign and the one guy started pandering to the audience, when you rolled your eyes and said, "What a suckup!"

If I'd said that about a presidential candidate when I was your age, I would have gotten it for disrespecting authority. So I had to ask myself, *Am I allowed to laugh at what she just said?* Answer: Yes. I agreed with you totally; he was groveling for the vote. I can't insist to you that all authority is worthy of your respect, because much of it is not. In five years you'll have to see through all the sucking-up and vote for your own president. Why *shouldn't* you start practicing now?

Every authority has its limits. I find myself defusing the menace of maleness by viewing it as a source of fascination. I study it constantly, not trying to learn how to *be* that, just trying to understand it. To say they run the world just doesn't cover it, because we do, too, in our less material way. Not in terms of real power, of course; it's impossible to imagine a reverse Saudi Arabia, in which we walked around doing whatever we pleased while forcing our entire male population to vacate themselves from public life and wear black cloth sacks with sideways slits for their eyes. We could never get them to do it; they're devoted to being in charge of things, and we seem unable to whip up any zeal for treating people like that. It's hard even to imagine a tradition of fine art in which naked men would recline on picnic blankets while fully clothed women looked on. Recently an artist in Colorado tried to communicate (especially to men, presumably) how it feels to have our sex so constantly and casually appropriated: She created a display of colorful penises pinned to a clothesline. The surfeit of masculine heebie-jeebies wrought by this little demonstration made national news, and lasted only days before a man broke in and destroyed the installation. I hope the artist has sense of humor enough to see that she made her point perfectly. Men rule, but in general seem to lack our fortitude.

And yet in some way or other their whole lives long, heterosexual guys are knocking themselves senseless to get our attention, and you can't help being charmed by the parade of nonsense. One of the most absurd, sexiest, most entrancing things I've ever seen took place right outside my study window. I was trying to think of a metaphor or something, staring out there into the mesquite woods, when suddenly my eyes snapped to focus on some movement: two rattlesnakes rising up together, face to face, as if they were being noodled up out of two snake charmers' baskets. Moving slowly with muscular, sinuous strength, they levitated nearly the entire front halves of their bodies, twisted themselves together, tussled a little, and finally slammed to the ground. It resembled arm wrestling. I ran to get everyone else in the house, and we all watched this thing go on for nearly an hour, the two snakes rearing up again and again, silently entwining, and then throwing themselves to the ground. We called our friend Cecil, the Arizona reptile expert, who informed us that arm wrestling wasn't such a bad analogy: These were two male snakes doing a dance of combat to win the favor of a female that was surely watching from somewhere nearby. We scanned the brush carefully from behind my window—these snakes were not even thirty feet away—and there she was, sure enough, stretched out languidly under a bush.

Then all at once, after innumerable tussles, according to some scoring system invisible to human eyes but unmistakable to the contestants, one guy won. The other slunk quickly away, and Sheba came sliding out into the open, with no eyelashes to bat but with love clearly on her mind, for off she slithered with her he-snake into the sunset. The greatest show on earth.

When you, my dear, were about two and a half, I carefully and honestly answered all the questions you'd started asking about reproductive organs. For several months thereafter, every time we met someone new, the unsuspecting adult would tousle your adorable blond head, and you'd look up earnestly and ask, "Do you have a penis or a vagina?"

If you are *ever* tempted to think my presence is an embarrassment to you, please recall that I stood by you during the "penis or vagina" months, July to September 1989. I wasn't sure I'd live through them or have any social life left afterward. I gave you a crash course in what we call "polite company" and harbored some doubts about whether honesty had really been the best policy.

What I see now, though, is that honesty *was*. Manners arrive in time; most girls are gifted enough at social savvy to learn the degree of polite evasion that will protect their safety and other people's dignity. But before anything else, you've got to be able to get the facts. Penis or vagina? I couldn't possibly tell you it wasn't to be discussed, or didn't matter. It matters, boy howdy, does it ever. Barbie or Ken, Adam or Eve, pilot or stewardess, knuckle sandwich or mea culpa, scissors, paper, rock, War and Peace. It's a very reasonable starting point. So begins the longest, scariest, sexiest, funniest, smartest, most extraordinary conversation we know. Cross your fingers, ready, set. Go.

Letter to My Mother

I imagine you putting on your glasses to read this letter. *Oh, Lord, what now?* You tilt your head back and hold the page away from you, your left hand flat on your chest, protecting your heart. "Dear Mom" at the top of a long, typed letter from me has so often meant trouble. Happy, uncomplicated things—these I could always toss you easily over the phone: I love you, where in the world is my birth certificate, what's in your zucchini casserole, happy birthday, this is our new phone number, we're having a baby in March, my plane comes in at seven, see you then, I love you.

The hard things went into letters. I started sending them from college, the kind of self-absorbed epistles that usually began as diary entries and should have stayed there. During those years I wore black boots from an army surplus store and a five-dollar haircut from a barbershop and went to some trouble to fill you in on the great freedom women could experience if only they would throw off the bondage of housewifely servitude. I made sideways remarks about how I couldn't imagine being anybody's wife. In my heart I believed that these letters—in which I tried to tell you how I'd become someone entirely different from the child you'd known—would somehow make us friends. But instead they only bought me a few quick gulps of air while I paced out the distance between us.

I lived past college, and so did my hair, and slowly I learned the womanly art of turning down the volume. But I still missed you, and from my torment those awful letters bloomed now and then. I kept trying; I'm trying still. But this time I want to say before anything else: Don't worry. Let your breath out. I won't hurt you anymore. We measure the distance in miles now, and I don't have to show you I'm far from where I started. Increasingly, that distance seems irrelevant. I want to tell you what I remember.

I'm three years old. You've left me for the first time with your mother while you and Daddy took a trip. Grandmama fed me cherries and showed me the secret of her hair: Five metal hairpins come out, and the everyday white coil drops in a silvery waterfall to the back of her knees. Her house smells like polished wooden stairs and soap and Granddad's onions and ice cream, and I would love to stay there always but I miss you bitterly without end. On the day of your return I'm standing in the driveway waiting when the station wagon pulls up. You jump out your side, my mother in happy red lipstick and red earrings, pushing back your dark hair from the shoulder of your white sleeveless blouse, turning so your red skirt swirls like a rose with the perfect promise of *you* emerging from the center. So beautiful. You raise one hand in a tranquil wave and move so slowly up the driveway that your body seems to be underwater. I understand with a shock that you are extremely happy. I have been miserable and alone waiting in the driveway, and you were at the beach with Daddy and *happy*. Happy without me.

I am sitting on your lap, and you are crying. *Thank you, honey, thank you,* you keep saying, rocking back and forth as you hold me in the kitchen chair. I've brought you flowers: the sweet peas you must have spent all spring trying to grow, training them up the trellis in the yard. You had nothing to work with but abundant gray rains and the patience of a young wife at home with pots and pans and small children, trying to create just one beautiful thing, something to take you outside our tiny white clapboard house on East Main. I never noticed until all at once they burst through the trellis in a pink red purple dazzle. A finger-painting of colors humming against the blue air: I could think of nothing but to bring it to you. I climbed up the wooden trellis and picked the flowers. Every one. They are gone already, wilting in my hand as you hold me close in the potato-smelling kitchen, and your tears are damp in my hair but you never say a single thing but *Thank you*.

Your mother is dead. She was alive, so thin that Granddad bought her a tiny dark-blue dress and called her his fashion model and then they all went to the hospital and came home without her. Where is the dark-blue dress now? I find myself wondering, until it comes to me that they probably buried her in it. It's under the ground with her. There are so many things I don't want to think about that I can't bear going to bed at night.

It's too hot to sleep. My long hair wraps around me, grasping like tentacles. My brother and sister and I have made up our beds on cots on the porch, where it's supposed to be cooler. They are breathing in careless sleep on either side of me, and I am under the dark cemetery ground with Grandmama. I am in the stars, desolate, searching out the end of the universe and time. I am trying to imagine how long forever is, because that is how long I will be dead for someday. I won't be able to stand so much time being nothing, thinking of nothing. I've spent many nights like this, fearing sleep. Hating being awake.

I get up, barefoot and almost nothing in my nightgown, and creep to your room. The door is open, and I see that you're awake, too, sitting up on the edge of your bed. I can make out only the white outline of your nightgown and your eyes. You're like a ghost.

Mama, I don't want to die.

You don't have to worry about that for a long, long time.

I know. But I'm thinking about it now.

I step toward you from the doorway, and you fold me into your arms. You are real, my mother in scent and substance, and I still fit perfectly in your lap.

You don't know what Heaven is like. It might be full of beautiful flowers.

When I close my eyes I discover it's there, an endless field of flowers. Columbines, blue asters, daisies, sweet peas, zinnias: one single flower bed stretching out for miles in every direction. I am small enough to watch the butterflies come. I know them from the pasture behind our house, the butterflies you taught me to love and name: monarchs, Dianas, tiger swallowtails. I follow their lazy zigzag as they visit every flower, as many flowers as there are stars in the universe. We stay there in the dark for a long time, you and I, both of us with our eyes closed, watching the butterflies drift so slowly, filling as much time as forever.

I will keep that field of flowers. It doesn't matter that I won't always believe in Heaven. I will suffer losses of faith, of love and confidence, I will have some bitter years, and always when I hurt and can't sleep I will close my eyes and wait for your butterflies to arrive.

Just one thing, I'm demanding of you. It's the middle of summer, humid beyond all reason, and I am thirteen: a tempest of skinned knees and menarche. You are trying to teach me how to do laundry, showing me how to put the bluing in with the sheets. The swampy Monday-afternoon smell of sheets drowning under the filmy, shifting water fills me with pure despair. I want no part of that smell. No future in white sheets and bluing. *Name one good thing about being a woman,* I say to you.

There are lots of good things. . . . Your voice trails off with the thin blue stream that trickles into the washer's indifferent maw.

In a rare flush of adrenaline or confidence, I hold on, daring you: *OK, then. If that's true, just name me one.*

You hesitated. I remember that. I saw a hairline crack in your claim of a homemaker's perfect contentment. Finally you said, *The love of a man. That's one thing. Being taken care of and loved by a man.*

And because you'd hesitated I knew I didn't have to believe it.

At fifteen I am raging at you in my diary, without courage or any real intention, yet, of actually revealing myself to you. *Why do you want to ruin my life? Why can't you believe I know how to make my own decisions? Why do you treat me like a child? No makeup or nail polish allowed in this house—you must think I am a baby or a nun. You tell me if I forget to close the curtains when I get undressed the neighbor boy will rape me. You think all boys are evil. You think if I go out with my girlfriends I'll get kidnaped. You think if I'm in the same room with a boy and a can of beer, I'll instantly become a pregnant alcoholic.*

Halfway through the page I crumble suddenly and write in a meeker hand, *I have to learn to keep my big mouth shut and not fight with Mom. I love her so much.*

I am a young woman sliced in two, half of me claiming to know everything and the other half just as sure I will never know anything at all. I am too awkward and quiet behind my curtain of waist-length hair, a girl unnoticed, a straight-A schoolmouse who can't pass for dumb and cute in a small-town, marry-young market that values—as far as I can see—no other type.

I understand this to be all your fault. You made me, and I was born a girl. You trained me to be a woman, and regarding that condition I fail to see one good thing.

The woodsmoke scent in the air puts me in mind of raked leaves, corduroy jumpers and new saddle shoes, our family's annual trip to Browning's orchard for apples and cider: a back-to-school nostalgia altogether too childish for me now, and yet here I am, thrilled to the edge of all my senses to be starting college. You and Dad have driven three hundred miles in our VW bus, which is packed like a tackle box with my important, ridiculous stuff, and now you have patiently unloaded it without questioning my judgment on a single cherished object—the plants, the turtle-shell collection, the glass demijohn, the huge striped pillow, the hundred books. You're sitting on my new bed while Dad carries in the last box. To you this bed must look sadly institutional compared with the furniture lovingly lathed for us from red cherrywood by your father before he died. To me the new metal bed frame looks just fine. Nothing fussy; it will do. I am arranging my plants in the windowsill while you tell me you're proud of the scholarship I won, you know I'll do well here and be happy, I should call if I need anything, call even if I don't.

I won't need anything, I tell you.

I am visited suddenly by a peculiar photographic awareness of the room, as if I were not really in it but instead watching us both from the doorway. I understand we are using up the very last minutes of something neither of us can call, outright, my childhood. I can't wait for you to leave, and then you do. I close the door and stand watching through my yellow-curtained window and the rust-orange boughs of a maple outside as you and Dad climb into the VW and drive away without looking back. And because no one can see me I wipe my slippery face with the back of my hand. My nose runs and I choke on tears, so many I'm afraid I will drown. I can't smell the leaves or apples or woodsmoke at all. I feel more alone than I've ever felt in my life.

At the Greencastle Drive-In on a double date, I am half of the couple in the backseat. We have the window open just a crack to accommodate the hissing metal speaker, and the heater is on full blast. Outside our happy island of steamy heat, frost is climbing the metal poles and whitening the upright bones of the surrounding cornfields; it's nearly Thanksgiving, surely the end of drive-in season as we know it. Tomorrow I have a midterm exam. Surprisingly, I don't even care. I feel heady and reckless. Truth be told, I will probably ace the exam, but even if I don't, so what? I'll live.

Finally I have a genuine social life and the privilege of giving in to peer pressure: I threw *Heart of Darkness* over my shoulder at the first bark of temptation and went out on a date.

The movie is *Cabaret.* Sally Bowles, with her weird haircut and huge, sad eyes, is singing her heart out about being abandoned by her mother.

With a physical shock I wake up to what's been tugging at my ear all day: November 20—it's your birthday! I've never forgotten your birthday, not since I was old enough to push a crayon around in the shape of a heart on a folded piece of construction paper. How can it be that this year I didn't even think to send a card? I sit bolt upright and open my mouth, preparing to announce that I have to go home immediately and call my mother. My friend and her boyfriend in the front seat are deeply involved in each other. I imagine them staring at me, hostile under rumpled hair, and feel myself shrinking into my former skin, the vessel of high school misery. I despise that schoolmouse. The happiness of my new adulthood is so precarious that I have to be careful not to wreck it like a ship. I sigh, settle back, shut my mouth, and watch Sally Bowles ruin her life.

I am nineteen, a grown woman curled like a fetus on my bed. Curled in a knot so small I hope I may disappear. I do not want to be alive.

I've been raped.

I know his name, his address—in fact I will probably have to see him again on campus. But I have nothing to report. Not to the police, not to you. The telephone rings and rings and I can't pick it up because it may be you. My mother. Everything you ever told me from the beginning has come home to this knot of nothingness on my bed, this thing I used to call me. I was supposed to prevent what happened. Two nights ago I talked to him at a bar. He bought me a drink and told my friends he thought I was cute. *That girl with the long hair,* he said. *What's her name?* Tonight when he came to my door I was happy, for ten full seconds. Then. My head against a wall, suffocation, hard pushing and flat on my back and screaming for air. Fighting an animal twice my size. My job was to stop him, and I failed. How can I tell you that? *You met him in a bar. You see?*

From this vantage point, a dot of nothingness in the center of my bed, I understand the vast ocean of work it is to be a woman among men, that universe of effort, futile whimpers against hard stones, and oh God I don't want it. My bones are weak. I am trapped in a room with no flowers, no light, a ceiling of lead so low I can never again straighten up. I don't want to live in this world.

I will be able to get up from this bed only if I can get up angry. Can you understand there is no other way? I have to be someone else. Not you, and not even me. Tomorrow or someday soon I will braid my long hair for the last time, go to my friend's house with a pair of sharp scissors, and tell her to cut it off. All of it. Tomorrow or someday soon I will feel that blade at my nape and the weight will fall.

Summer light, Beaurieux, France. At twenty-three I'm living in an enormous, centuries-old stone farmhouse with a dozen friends, talkative French socialists and a few British expatriates, all of us at some loose coupling in our lives between school and adulthood. We find a daily, happy solace in one another and in the scarlet poppies that keep blooming in the sugar-beet fields. We go out to work together in the morning and then come home in the evening to drink red table wine and make ratatouille in the cavernous stone-floored kitchen. This afternoon, a Saturday, a gaggle of us have driven into town to hang out at the village's only electrified establishment, a tiny café. We are entrenched in a happy, pointless argument about Camus when the man wiping the counter answers the phone and yells, *"Mademoiselle Kingsolver? Quelque'un des Etats Unis!"*

My heart thumps to a complete stop. Nobody from the United States can possibly know where I am. I haven't written to my parents for many months, since before I moved to France. I rise and sleepwalk to the telephone, knowing absolutely that it will be you, my *mother,* and it is. I still have no idea how you found me; I can hardly even remember our conversation. I must have told you I was alive and well, still had all my arms and legs—what else was there for me to say? You told me that my brother was getting married and I ought to come home for the wedding.

Ma mere! This is what I tell my friends, with a shrug, when I return to the table. I tell them you probably called out the French Foreign Legion to find me. Everybody laughs and declares that mothers are all alike: They love us too much, they are a cross to bear, they all ought to find their own things to do and leave us alone. We pay for our coffee and amble toward somebody's car, but I decide on impulse I'll walk back to the farm. I move slowly, turning over and over in my mind the telephone's ringing, the call that was for me. In France, a tiny town, the only café, a speck of dust on the globe. Already it seems impossible that this really happened. The roadside ditch is brilliant with poppies, and as I walk along I am hugging myself so hard I can barely breathe.

I'm weeding the garden. You admired my garden a lot when you came to visit me here in Tucson, in this small brick house of my own. We fought, of course. You didn't like my involvement with Central American refugees, no matter how I tried to talk you through the issues of human rights and our government's support of a dictatorship in El Salvador. Even more, you didn't like it that I was living in this little brick house with a man, unmarried. But you did admire my vegetable garden, and the four-o'clocks in the front yard, you said, were beautiful. The flowers were our common ground. They'll attract hummingbirds, you told me, and we both liked that.

On this day I am alone, weeding the garden, and a stranger comes to the door. He doesn't look well, and he says he needs a glass of water, so I go to the kitchen. When I turn around from the sink, there he is, with a knife shoved right up against my belly.

Don't scream or I'll kill you.

But I do scream. Scream, slap, bite, kick, shove my knee into his stomach. I don't know what will happen next, but I know this much: It's not going to be my fault. You were partly wrong and partly right; bad things are bound to happen, but this hateful supremacy that sometimes shoves itself against me is *not* my fault. You've held on to me this long, so I must be someone worth saving against the odds. When I can finally tell you that, I know you will agree with me. So this time I scream. I scream for all I'm worth.

I'm somewhere between thirty and a hundred, and I've written a book. My first novel, but to me it seems more like the longest letter to you I've ever written. Finally, after a thousand tries, I've explained everything I believe in, exactly the way I always wanted to: human rights, Central American refugees, the Problem That Has No Name, abuse of the powerless, racism, poetry, freedom, childhood, motherhood, Sisterhood Is Powerful. All that, and still some publisher has decided it makes a good story.

But that doesn't matter right now, because you are on my

front-porch glider turning the pages one at a time, and it's only a stack of paper. The longest typed letter ever, a fistful of sweet peas, the world's largest pile of crayoned hearts. I'm supposed to be cooking dinner but I keep looking out the window, trying to see what page you're on now, trying to read your thoughts from the back of your head. All I can see is that you're still reading, and your hair is gray. In another few minutes mine will be, too. My heart is pounding. I boil water and peel the potatoes but forget to put them in the pot. I salt the water twice. At last you come in with tears in your eyes.

Barb, honey, it's beautiful. So good.

You fold me into your arms, and I can't believe it: I still fit. Or I fit once again.

I'm thirty-two, with my own daughter in my arms. I've sent you a picture of her, perfect and gorgeous in her bassinette. Her tiny hand is making a delicate circle, index finger to thumb, pinkie extended as if she were holding a teacup. How could my ferocious will create such a delicate, feminine child? *This one is all girl,* I write on the back, my daughter's first caption. You send back a photo of me at the same age, eight weeks, in my bassinette. I can't believe it: I am making a delicate circle with my hand, index finger to thumb, pinkie extended.

We've flown back to Kentucky for Christmas. I'm thirty-seven but feeling weirdly transported backward. For all the years I've been away from my hometown, Main Street is perfectly unchanged. The commercial district is set off by a stoplight on either end of a single block: hardware store, men's clothing, five-and-dime, drugstore, county jail. An American flag made of red, white, and blue light bulbs (some of them burned out) blinks to life on the courthouse's peeling silver dome at dusk. Who on earth was I, in the years I called this place home?

I've come into town to run a few errands, and to tell the truth, it's a relief to take a few hours' respite from a house overstuffed with Christmas decorations and sugary fruitcakes, a kitchen stifling from an overworked oven, and hugs at every turn. How did I ever live *there?* My own life is so different now. So pared down. In fact, I haven't found a way to tell you this, but my marriage is slowly dying, and I will soon be on my own again—this time next year I'll be a single mother. I am handling this, I cope. I stop in at Hopkins' Drug to get something for my daughter's cough. The bell over the door jingles, and I wave a leather-gloved hand at Mr. Hopkins, still filling prescriptions at the back of the store. I tell him yes, I'm home visiting, and ask what he recommends for my daughter. She picked up a cold on the trip; she hasn't been sick many times in her life, so I'm worried, but only a little. In my tweed winter coat and convenient shoulder-length haircut I feel competent and slightly rushed, as usual. A woman of my age.

An elderly woman I don't recognize stops me with an arthritic hand across my wrist as I'm about to leave the drugstore. Her eyes swim toward me like dark fish as she eyes me through thick glasses, up and down. *You must be Virginia Kingsolver's daughter. You look exactly like her.*

My nine-year-old daughter comes home from a summer slumber party with painted nails, and I mean *painted.* Day-Glo green on the fingers, purple on the toes. We drive to the drugstore for nail-polish remover.

Please! All the girls my age are doing this.

How can every nine-year-old on the planet possibly be painting her toenails purple?

I don't know. They just are.

School starts in a week. Do you want to be known by your teacher as the girl with the green fingernails?

Yes. But I guess you *don't.*

Do you really?

She looks down at her nails and states: *Yes.* With her porcelain skin and long, dark lashes, she is a Raphael cherub. Her perfect mouth longs to pout, but she resists, holds her back straight. A worthy vessel for her own opinions. Despite myself, I admire her.

OK, we'll compromise. The green comes off. But keep the purple toenails.

At forty, I'm expecting my second child. Through my years of being coupled and then alone, years of accepting my fate and then the astonishing chance of remarriage, I've waited a lifetime for this gift: a second child. But now it is past due, and I am impatient. I conceived in late September and now it's July. I have dragged this child in my belly through some portion of every month but August. In the summer's awful heat I am a beached whale, a house full of water, a universe with ankles. It seems entirely possible to me that the calendar will close and I will somehow be bound fairy-tale-wise to a permanent state of pregnancy.

During these months you and I have talked more often than ever before. Through our long phone conversations I've learned so many things: that you fought for natural childbirth all three times, a rebel against those patronizing doctors who routinely knocked women out with drugs. In the fifties, formula was said to be modern and breast-feeding crude and old-fashioned, but you ignored the wagging fingers and did what you and I both know was best for your babies.

I've also learned that ten-month pregnancies run in our family.

When your sister was two weeks overdue, I made Wendell drive over every bumpy road in the county.

Did it work?

No. With you, we were in Maryland. We drove over to see the cherry blossoms on the Capitol Mall in April, and Wendell said, "What if you go into labor while we're hours away from home?" I told him, "I'll sing Hallelujah."

A week past my due date you are calling every day. Steven answers the phone, holds it up, and mouths, "Your mother again." He thinks you may be bugging me. You aren't. I am a woman lost in the weary sea of waiting, and you are the only one who really knows where I am. Your voice is keeping me afloat. I grab the phone.

She is born at last. A second daughter. I cry on the phone, I'm so happy and relieved to have good news for you, finally. I promise you we'll send pictures right away. You will tell me she looks just like I did. She looks like her father, but I will believe you anyway.

Later on when it's quiet I nurse our baby, admiring her perfect hands. Steven is in a chair across the room, and I'm startled to look up and find he is staring at us with tears in his eyes. I've seen him cry only once before.

What's wrong?

Nothing. I'm just so happy.

I love him inordinately. I could not bear to be anyone but his wife just now. I could not bear to be anyone but the mother of my daughters.

I was three years old, standing in the driveway waiting for the car to bring you back from Florida. You arrived glowing with happiness. *Because of me.* I felt stung, thinking you could carry on your life of bright-red lipstick smiles outside of my presence, but I know now I was wrong. You looked happy *because of me.* You hadn't seen me for more than a week, hadn't nursed me for years, and yet your breasts tingled before you opened the car door. The soles of your feet made contact with the ground, and your arms opened up as you walked surefooted once again into the life you knew as my mother. I know exactly how you felt. I am your happiness. It's a cross I am willing to bear.

Going to Japan

My great-aunt Zelda went to Japan and took an abacus, a bathysphere, a conundrum, a diatribe, an eggplant. That was a game we used to play. All you had to do was remember everything in alphabetical order. Right up to Aunt Zelda.

Then I grew up and was actually invited to go to Japan, not with the fantastic Aunt Zelda but as myself. As such, I had no idea what to take. I knew what I planned to be doing: researching a story about the memorial at Hiroshima; visiting friends; trying not to get lost in a place where I couldn't even read the street signs. Times being what they were—*any* times—I intended to do my very best to respect cultural differences, avoid sensitive topics I might not comprehend, and, in short, be anything but an Ugly American. When I travel, I like to try to blend in. I've generally found it helps to be prepared. So I asked around, and was warned to expect a surprisingly modern place.

My great-aunt Zelda went to Japan and took Appliances, Battery packs, Cellular technology. . . . That seemed to be the idea.

And so it came to pass that I arrived in Kyoto an utter foreigner, unprepared. It's true that there are electric streetcars there, and space-age gas stations with uniformed attendants who rush to help you from all directions at once. There are also golden pagodas on shimmering lakes, and Shinto shrines in the forests. There are bamboo groves and nightingales. And finally there are more invisible guidelines for politeness than I could fathom. When I stepped on a streetcar, a full head taller than all the other passengers, I became an awkward giant. I took up too much space. I blended in like Igor would blend in with the corps de ballet in *Swan Lake*. I bumped into people. I crossed my arms when I listened, which turns out to be, in Japanese body language, the sign for indicating brazenly that one is bored.

But I wasn't! I was struggling through my days and nights in the grip of boredom's opposite—i.e., panic. I didn't know how to eat noodle soup with chopsticks, and I did it most picturesquely *wrong*. I didn't know how to order, so I politely deferred to my hosts and more than once was served a cuisine with heads, including eyeballs. I managed to wrestle these creatures to my lips with chopsticks, but it was already too late by the time I got the message that *one does not spit out anything.*

I undertook this trip in high summer, when it is surprisingly humid and warm in southern Japan. I never imagined that in such sweltering heat women would be expected to wear stockings, but every woman in Kyoto wore nylon stockings. Coeds in shorts *on the tennis court* wore nylon stockings. I had packed only skirts and sandals; people averted their eyes.

When I went to Japan I took my Altitude, my Bare-naked legs, my Callous foreign ways. I was mortified.

My hosts explained to me that the Japanese language does not accommodate insults, only infinite degrees of apology. I quickly memorized an urgent one, *"Sumimasen,"* and another for especially extreme cases, *"Moshi wake gozaimasen."* This translates approximately to mean, "If you please, my transgression is so inexcusable that I wish I were dead."

I needed these words. When I touched the outside surface of a palace wall, curious to know what it was made of, I set off screeching alarms and a police car came scooting up the lawn's discreet gravel path. "*Moshi wake gozaimasen,* Officer! Wish I were dead!" And in the public bath, try as I might, I couldn't get the hang of showering with a hand-held nozzle while sitting fourteen inches from a stranger. I sprayed my elderly neighbor with cold water. In the face.

"Moshi wake gozaimasen," I declared, with feeling.

She merely stared, dismayed by the foreign menace.

I visited a Japanese friend, and in her small, perfect house I spewed out my misery. "Everything I do is wrong!" I wailed like a child. "I'm a blight on your country."

"Oh, no," she said calmly. "To forgive, for us, is the highest satisfaction. To forgive a foreigner, ah! Even better." She smiled. "You have probably made many people happy here."

To stomp about the world ignoring cultural differences is arrogant, to be sure, but perhaps there is another kind of arrogance in the presumption that we may ever really build a faultless bridge from one shore to another, or even know where the mist has ceded to landfall. When I finally arrived at Ground Zero in Hiroshima, I stood speechless. What I found there was a vast and exquisitely silent monument to forgiveness. I was moved beyond words, even beyond tears, to think of all that can be lost or gained in the gulf between any act of will and its consequences. In the course of every failure of understanding, we have so much to learn.

I remembered my Japanese friend's insistence on forgiveness as the highest satisfaction, and I understood it really for the first time: What a rich wisdom it would be, and how much more bountiful a harvest, to gain pleasure not from achieving personal perfection but from understanding the inevitability of imperfection and pardoning those who also fall short of it.

I have walked among men and made mistakes without number. When I went to Japan I took my Abject goodwill, my Baleful excuses, my Cringing remorse. I couldn't remember everything, could not even recite the proper alphabet. So I gave myself away instead, evidently as a kind of public service. I prepared to return home feeling empty-handed.

At the Osaka Airport I sat in my plane on the runway, waiting to leave for terra cognita, as the aircraft's steel walls were buffeted by the sleet and winds of a typhoon. We waited for an hour, then longer, with no official word from the cockpit, and then suddenly our flight was canceled. Air traffic control in Tokyo had been struck by lightning; no flights possible until the following day.

"We are so sorry," the pilot told us. "You will be taken to a hotel, fed, and brought back here for your flight tomorrow."

As we passengers rose slowly and disembarked, we were met by an airline official who had been posted in the exit port for the sole purpose of saying to each and every one of us, "Terrible, terrible. *Sumimasen*." Other travelers nodded indifferently, but not me. I took the startled gentleman by the hands and practically kissed him.

"You have no idea," I told him, "how thoroughly I forgive you."

Life Is Precious, or It's Not

Columbine used to be one of my favorite flowers," my friend told me, and we both fell silent. We'd been talking about what she might plant on the steep bank at the foot of the woods above her house, but a single word cut us suddenly adrift from our focus on the uncomplicated life in which flowers could matter. I understood why she no longer had the heart to plant columbines. I feel that way, too, and at the same time I feel we ought to plant them everywhere, to make sure we remember. In our backyards, on the graves of the children lost, even on the graves of the children who murdered, whose parents must surely live with the deepest emotional pain it is possible to bear.

In the aftermath of the Columbine High School shootings in Colorado, the whole country experienced grief and shock and—very noticeably—the spectacle of a nation acting bewildered. Even the op-ed commentators who usually tell us just what to think were asking, instead, what we should think. How could this happen in an ordinary school, an ordinary neighborhood? Why would any student, however frustrated with meanspirited tormentors, believe that guns and bombs were the answer?

I'm inclined to think all of us who are really interested in these questions might have started asking them a long while ago. Why does any person or nation, including ours, persist in celebrating violence as an honorable expression of disapproval? In, let's say, Iraq, the Sudan, Waco—anywhere we get fed up with meanspirited tormentors—why are we so quick to assume that guns and bombs are the answer?

Some accidents and tragedies and bizarre twists of fate are truly senseless, as random as lightning bolts out of the blue. But this one at Columbine High was not, and to say it was is irresponsible. "Senseless" sounds like "without cause," and it requires no action, so that after an appropriate interval of dismayed handwringing, we can go back to business as usual. What takes guts is to own up: This event made sense. Children model the behavior of adults, on whatever scale is available to them. Ours are growing up in a nation whose most important, influential men—from presidents to the coolest film characters—solve problems by killing people. Killing is quick and sure and altogether manly.

It is utterly predictable that some boys who are desperate for admiration and influence will reach for guns and bombs. And it's not surprising that this happened in a middle-class neighborhood; institutional violence is right at home in the suburbs. Don't let's point too hard at the gangsta rap in our brother's house until we've examined the video games, movies, and political choices we support in our own. The tragedy in Littleton grew out of a culture that is loudly and proudly rooting for the global shootout. That culture is us.

Conventional wisdom tells us that Nazis, the U.S. Marines, the Terminator, and the NYPD. all kill for different reasons. But as every parent knows, children are good at ignoring or seeing straight through the subtleties we spin. Here's what they must surely see: Killing is an exalted tool for punishment and control. Americans who won't support it are ridiculed, shamed, or even threatened. The Vietnam war was a morally equivocal conflict by any historical measure, and yet to this day, candidates for public office who avoided being drafted into that war are widely held to be unfit for leadership.

Most Americans believe bloodshed is necessary for preserving our way of life, even though it means risking the occasional misfire—the civilians strafed because they happened to live too close to the terrorist, maybe, or the factory that actually made medicines but *might* have been making weapons. We're willing to sacrifice the innocent man condemned to death row because every crime must be paid for, and no jury is perfect. The majority position in our country seems to be that violence is an appropriate means to power, and that the loss of certain innocents along the way is the sad but inevitable cost.

I'd like to ask those who favor this position if they would be willing to go to Littleton and explain to some mothers what constitutes an acceptable risk. Really. Because in a society that embraces violence, this is what "our way of life" has come to mean. The question can't be *why* but only "Why yours and not mine?" We have taught our children in a thousand ways, sometimes with flag-waving and sometimes with a laugh track, that the bad guy deserves to die. But we easily forget a crucial component of this formula: "Bad" is defined by the aggressor. Any of our children may someday be, in someone's mind, the bad guy.

For all of us who are clamoring for meaning, aching for the loss of these precious young lives in Littleton to mean something, my strongest instinct is to use the event to nail a permanent benchmark into our hearts: Life is that precious, period. It is possible to establish zero tolerance for murder as a solution to anything. Those of us who agree to this contract can start by removing from our households and lives every television program, video game, film, book, toy, and CD that presents the killing of humans (however symbolic) as an entertainment option, rather than the appalling loss it really is. Then we can move on to harder choices, such as discussing the moral lessons of capital punishment. Demanding from our elected officials the subtleties and intelligence of diplomacy instead of an endless war budget. Looking into what we did (and are still doing) to the living souls of Iraq, if we can bear it. And—this is important—telling our kids we aren't necessarily proud of the parts of our history that involved bombing people in countries whose policies we didn't agree with.

Sounds extreme? Let's be honest. *Death* is extreme, and the children are paying attention.

Flying

I've traveled in airplanes so often, I have frequent-flyer miles enough to go to China.

I never watch movies that feature violent, spectacular horrors, so my uncalloused psyche was laid wide open to the images, when they came, of real airplanes slamming into real buildings.

And I have an overgrown, acutely visual imagination. It's the combination of these vulnerabilities, I suppose, that nearly debilitated me for many days with my own particular visions of what hundreds of people must have gone through—but did not live through—on Tuesday, September 11, 2001.

In the weeks following that monstrous massacre I walked through the motions of a normal life, like everyone else who was lucky enough to have a "normal" to get back to, rather than an aching hole where a loved one used to be. In literal terms I was untouched; my home is thousands of miles from any site where an airplane crashed that day. But I have many friends who are much closer to the catastrophe: one who often works near the Pentagon; another who was passing through Washington Square on his way to work in lower Manhattan when his eyes went up to the tallest building as it began, incomprehensibly, to fall down. He likened the sound that rose from the bystanders to "a packed stadium filled with keening." One of the people dearest to me on earth was on a plane that took off from Newark, I was certain, at the same hour and minute as two of the fatal flights. For the first time in my life, my calls to that city that never sleeps were answered with a dead line. I was worried sick about my far-flung friends for the hours and days it took until I could talk with each one in turn, reassuring myself that my community remained more or less as it had been. I have two close friends who lost people they loved, so I stand one degree separated from a tragedy that directly bereaved so many in one horrendous blow.

Only my soul was scathed. My mind's eye kept watching this movie in my head: seeing the blue stars that invade my vision in times of panic; breathing too fast, gripping the hand of a stranger in the seat beside me; thinking in frantic minutes about the years my girls will have to get through without me. Wishing I *had* after all, despite my grudge against the things, bought a cell phone. Brutal murder with knives; desperation; watching the end of the world from an aisle seat.

So many people. They crowded my consciousness with a silent cry for memorial. I woke from dreams of facing the end beside them, and in those first confused seconds while I struggled to identify the impalpable burden that weighed on my heart I would see it again, not as dream but as reality. Throughout my day I would find myself gripped by distraction, looking out the window at the haze on the mountains as the scene played in front of me again and again. Only a few days before September 11, I had needed to undertake the new (for me) experiences of general anesthesia and surgery. And so the acute pain, foggy-headedness, and disturbed sleep of my slow recuperation became confused and intermingled for me with the pain and recuperative trauma of my nation in a slow, hallucinatory grief.

Then a death in our own family followed upon all those others by just three days, a further devastation, and we faced the difficulties of trying to get to my father-in-law's funeral in a distant city. The airways reopened, but a timely arrival was by no means guaranteed. I wasn't yet well enough to travel, so I stayed home with the children while Steven made the trip alone. As he left, I wept again from my bottomless well of grief, feeling sure the world was ending—feeling sure that my husband's airplane would also fall out of the sky, and I would not see him again.

"It's probably safer to fly right now," he told me reasonably, "than at any time in the last ten years."

"I know that," I said, "but I don't *feel* that."

This is what changed for us that day: not what we know, but how we feel. We have always lived in a world of constant sorrow and calamity, but most of us have never had to say before, It could have been me. My daughters and me on that plane, my husband in that building. I have stepped on that very pavement, I have probably sat on one of those planes. This was *us*, Americans at work, on vacation, going home, or just walking from one building to another. Alive, then dead.

It's probably only human to admit that a stranger's death is more shattering when we can imagine it as our own. We all began to say, that week, "This is the worst thing that has ever happened." *To us,* I know we should have added, for worse disasters have happened—if "worse" can be measured solely by the number of dead—in practically every other country on earth. Two years earlier an earthquake in Turkey had killed three times as many people in one day, babies and mothers and businessmen. The November before that, a hurricane had hit Honduras and Nicaragua and killed even more, buried whole villages and erased family lines; even now, people wake up there empty-handed. Some disasters are termed "natural" (though it was war that left Nicaragua so vulnerable), and yet their victims are just as innocent as ours on September 11, and equally dead. Which end of the world should we talk about? Only the murderous kind? Sixty years ago, Japanese airplanes bombed U.S. Navy boys who were sleeping on ships in gentle Pacific waters. Three and a half years later, American planes bombed a plaza in Japan where men and women were going to work and schoolchildren were playing, and more humans died at once than anyone had ever thought possible: seventy thousand in a minute. Imagine, now that we can—now that we have a number with which to compare it—*seventy thousand people dead in one minute.* Then twice that many more, slowly, from the inside.

There are no worst days, it seems. Ten years ago, early on a January morning, bombs rained down from the sky and caused great buildings in the city of Baghdad to fall down—hotels, hospitals, palaces, buildings with mothers and soldiers inside—and here in the place I want to love best, I watched people cheer about it. In Baghdad, survivors shook their fists at the sky and used the word *evil.* We all tend to raise up our compatriots' lives to a sacred level, thinking our own citizens to be more worthy of grief and less acceptably taken than lives on other soil. When many lives are lost all at once, people come together and speak words such as *heinous* and *honor* and *revenge,* presuming to make this awful moment stand apart somehow from the ways people die a little each day around the world from sickness or hunger. But broken hearts are not mended in this ceremony because really, every life that ends is utterly its own event—even as in some way every life is the same as all others, a light going out that ached to burn longer. Even if you never had a chance to love the light that's gone, you miss it. You should; you have to. You bear this world and everything that's wrong with it by holding life still precious, every time, and starting over.

In my lifetime I have argued against genocide, joined campaigns for disaster aid, sent seeds to places of famine. I have mourned my fellow humans in every way I've known how. But never before have their specific deaths so persistently entered my dreams. This time it was *us,* leaving us trembling, leading my little daughter to ask quietly, "Will it happen to me, Mama?" I understood with the deepest sadness I've ever known that this was the wrong question to ask, and it always had been. It has *always* been happening to us—in Nicaragua, in the Sudan, in Hiroshima, that night in Baghdad—and now we finally know what it feels like. Now we may learn, from the taste of our own blood, that every war is both won and lost, and that loss is a pure, high note of anguish like a mother singing to an empty bed.

In the coming spring I will plant a long raft of Legion of Honor poppies, the same ones that bloom in the graveyards at Verdun, across the bottom of our hay field. I decided to do this early on, as my own cenotaph. Every summer when the poppies' scarlet heads rise up from the earth, I'll remember that grief is eternal and so is life.

So many people were taken from us all at once this year, such courage and grief and fury cry out silently through our still-beating hearts, asking our nation for the right memorial. How can we build it, what material shall we choose? What will be the quality of its soul? I've seen the thousand paper cranes in the quiet monument to peace in Hiroshima; the dark slash in the earth whispering the names of those we lost in Vietnam; the endless orchard of white crosses in the poppy fields of Verdun. And I've studied how all these nations behaved in the aftermath of their losses. The most crucial memorial must surely be in what we carry forward.

It will probably always be hard to speak of September 11 without using the words *unthinkable, unbelievable.* And yet somehow we did and do believe it. From the first moment I understood what was happening to us that morning, I felt my bones going soft with the most awful recognition I've ever known, the aching perception that this had been working its way toward us for so long. I did not, and will never, believe that such a blow was deserved; no one *deserves* to be murdered, least of all those blameless people sitting down to their desks, carrying breakfast to diners in a restaurant, or setting off on vacation. What I do believe is that the losers of all wars are largely the innocent, and we are a nation at war—we have been for many lifetimes, reinforcing or inventing reigns of power that mollify some and terrify others in many lands, for many reasons. In the years since I became a taxpaying adult, my money has helped finance air assaults in Afghanistan, Chile, El Salvador, Grenada, Iran, Iraq, Libya, Lebanon, Nicaragua, Panama, the Sudan, Vietnam, and Yugoslavia—and those are just the ones I can name without getting up from my desk to search out the history books that will remind me of all the others. How could I—how could *anyone*—reasonably have expected that we could go about our merry lives here in War Headquarters without being touched by war? Whether or not we deem all these campaigns just, we can't possibly expect to wage war without living in wartime. Elsewhere on the planet, no one is banking on that program. War is not just some game played by the strong against, or on behalf of, the weak. Lethal germs, airliner bombs, nuclear sabotage, suicide missions—these are the weapons of the fiercely resourceful warrior raging against the mighty. In my bones I understand that they are not just going to be aimed at others *every* time; at some point they will be coming toward me. Assigning blame on this score is about as useful as looking at a hurricane through a drinking straw. The vast injuries comprising this picture surely call for justice, but the word *justice* itself calls for a system of accounting that may not be up to the task, when so many wounds have been inflicted for different reasons over the course of centuries, all in the guise of retaliation. I feel intensely linked with those who died on September 11, not because of any particular culpability, and not because I can presume to take anyone's place in any story, but because their deaths forced me wide awake to the immensity of what is wrong. This story belongs to every one of us, as a price somehow exacted in blood and anger and conscience. I am linked to a chain of events that has wrought devastation.

In those first weeks I found myself not stunned by this truth but in some way lost, watching the end of the world on an airplane again and again. I needed to finish this out so the awful loop in my head might finally close. I needed to understand what the dead were asking of me. People often say funerals and memorials are for the living, not the dead, and maybe that is so. But if we can't summon the empathy to imagine what our dead would have asked of us, or the selflessness to give it, then we must accept the desperately sad verdict that each generation's hopes will die with it, and no cumulative progress is possible for the human will. Thousands of people awakened on the morning of September 11 and hurried or lingered over coffee, saw children off to school, made plans, kissed spouses, and bent their shoulders to the good they hoped to make of a new day, without ever dreaming they would not see the sun set again. To transsubstantiate these unfinished lives into some honorable endeavor on behalf of what they might have been—this is surely more necessary now than satisfying our own anger and cravings for safety.

I can only hope to know what is being asked of me if I can somehow get beyond a survivor's panic and anger, to fly through all this or above it to glimpse another point of view. Possibly this is what each of us needs in order to go forward, the answer to the hardest question: If it could have been me on that awful Tuesday—if it could still be me, and the ones closest to me, the next time—what memorial would I ask the world to build for our remembrance? What would I want to have written on stone? Not to complete some national agenda, but just for *us?*

First what goes without saying and has to be said: I would want my husband and children to know I loved them. But they know this every minute of our lives, from the way I have always lived—for them, with them, in every way cheering them on; *with* them even when I've had to be apart from them. Every survivor must seize this comfort. And I would hope all members of my family—including those who are unrelated by blood but who've proven themselves to be my devoted tribe in good times and bad—would remember what fine people they have been in my eyes, and carry that reinforcement into the rest of their days. That is the best reincarnation I can imagine: to be a cricket on someone's hearth, a small, encouraging light in the heart of a friend who looks at the world and its challenges as I would.

And my daughters, as precious as my eyes: I would have them be brave enough, and gentle enough, to remember me by embracing the world and engaging in its design. I wouldn't need to know how they'd do it, only that they would earn the unquenchable happiness that comes to those who leave a place more beautiful, somehow, for their having walked through it.

I would want someone to plant a raft of poppies for me.

It is one thing to imagine the gentle end that most of us envision for ourselves, with our survivors at our bedside; to imagine death by violence, and its aftermath, demands a magnitude of labor I can scarcely marshal. A murder is an unspeakable thing to have to pack away in a human heart. It brings the temptation to bitterness, a particular pain that cries out to be healed with pain. There comes a momentary sense of conviction: that some other's anguish will erase our own. I have felt that rage myself, when I was attacked by someone who did not care if I lived or died; I've felt it even more fiercely on the occasion of harm, real or threatened, to my children. I still sometimes feel an intense longing, as sharp as the knife that once touched my ribs, to hurt another person. I felt it, of course, after September 11. Even now the feeling rises like the undead, adapts like a germ to new bodies and forms, tries to pass itself off as righteous at times. I struggle to forgive the unforgivable and doubt I am up to the task.

But I have lived long enough and had the help of wise enough friends to feel how the stones in my heart can settle when the substrate of my rage transforms into a kinder species of force. Sometimes I've survived anger only one minute at a time, by saying to myself again and again that the best revenge is some kind of life beyond this, some kind of goodness. And I can lay no claim to goodness until I can prove that mean people have not made me mean.

This transformation isn't easy; possibly nothing is harder. In the days after September 11, my mind led me repeatedly to the edge of tasting my own death at the hands of a heartless murderer, though never into the more awful terrain, of which I can hardly make myself speak, of surviving the loss of a child. I was brought to tears during those weeks by the expansive grace of some of the parents who did lose sons and daughters on those airplanes and in those buildings. Oscar Rodriguez, whose son died in the World Trade Center, said, "I know there is anger. I feel it myself. But I don't want my son used as a pawn to justify the killing of others. We as a nation should not use the same means as the people who attacked us." And a bereaved mother in Washington, D.C., was able to assemble the composure, only days after her daughter's murder, to speak publicly of her gratitude for the years she had been given with her twelve-year-old, rather than reviling the men who had stolen the many more years she might have had. Rita Lasar, the sister of a man who died on the twenty-seventh floor of one of the towers because he stayed with a wheelchair-bound coworker while others fled, brought me to know her grief when she wrote of having helped raise her much younger, much-beloved brother. And she brought me to understand the magnitude of her love when she declared eloquently, as our country rushed toward war, "I will stay behind, just as my dear brother Avrame did. I will stay behind and ask America not to do something we can't take back." These loving family members are a monument. To take them at their word is to understand what is possible from the human spirit.

There can be no greater spiritual accomplishment than to come through brutal trials and then look back and see that mean times did not render us mean spirits. If I had never been granted the chance to do this, I would still want it for my children. Of all the fates I can imagine for myself, no legacy leaves me colder than that of bitterness and hatred. I would rather be forgotten entirely than held in any way responsible for the vengeful loss of a single life, let alone thousands of lives, or any historic moment of jingoism or ethnic hatred. I feel chilled and forsaken when I picture kissing my children good-bye some morning and, by nightfall, having all the beauty of my days reduced to a symbol claimed by military men as an act of war. No bomb has ever been built that can extinguish hatred, and while I have been told that this is not the point, I insist on it as *my* point, if one is ever to be made for me. Vengeance does not subtract any numbers from the equation of murder; it only adds them. Empathy, comprehension, resolution—these are the only powers for murder's reduction. Who would really wish to be a cause for taking bread from the mouths of the hungry in a desperate rush to beat more of the world's plowshares into swords? That is surely a despicable memorial. I speak for many more than myself, I know, when I declare that I'd rather be remembered as a lesson learned, a sympathy made acute, a moment in which humanity rose humanely to a fearful occasion.

A close friend of mine in Virginia who is also a mother and writer, with whom I have shared the struggle against the culturally complex anachronisms of Dixie, wrote me a few days after the fall with this report: "Even the Confederate flags are at half mast." That sentence struck me as a poem, complete in itself.

Maybe we've had the wind taken out of our sails for just long enough that our course will realign, however infinitesimally, toward a kinder star. We would not be making a concession to murder or its perpetrators if we were to learn from our fears and our losses. Martin Luther King Jr., four little girls from Selma, and hundreds of other murdered souls in our history have given us a pause in which to examine the national conscience and embrace a more generous vision of ourselves than we ever thought possible. That must be our monument to the lost.

I've lived long enough to eat many youthful words, but a few things I have always known for certain, and this is one: If I had to give up my life for anything, it would have to have the resilience of hope, the elation of new literacy, the brilliant life of a field of flowers, the elementary kindness of bread. Nothing short of that. It would have to be something as sure as love.

Household Words

I was headed home with my mind on things; I can't even say what they were. It was an afternoon not very long ago, and probably I was ticking through the routine sacrament of my day—locating every member of my family at that moment and organizing how we would all come together for dinner and what I would feed us—when my thoughts were bluntly interrupted. A woman was being attacked fifty feet away from me. My heart thumped and then seemed to stop for good and then thumped hard again as I watched what was

happening. The woman was slight, probably no taller than my older daughter, but she was my age. Her attacker, a much taller man, had no weapon but was hitting her on the head and face with his fists and open hands, screaming, calling her vile names right out in the open. She ducked, in the way any animal would, to save the more fragile bones of her face. She tried to turn her back on him, but he pursued her, smacking at her relentlessly with the flat of his hand and shouting angrily that she was trash, she was nothing, she should get away from him. And she was trying, but she couldn't. I felt my body freeze as they approached. They came very close, maybe ten feet away or even less, and then they moved on past us. I say *us* because I wasn't alone here: I was in a crowd of several dozen people, all within earshot. Maybe there were closer to a hundred of us; I'm not sure. Unbelievably, most weren't even looking. And then I did my own unbelievable thing: I left. I moved forward toward my home and family and left that battered woman behind.

I did and I didn't leave her behind, because I'm still thinking and now writing about this scene, reviling my own cowardice. Reader, can you believe I did what I did? Does it seem certain that I am heartless?

Let me give some more details of the scene, not because I hope to be forgiven. I ask only that all of us try to find ourselves in this weird landscape. It was the United States of America. I was at a busy intersection, in a car. The woman had the leathery, lined face and tattered-looking hair of a person who lives her whole life outdoors beneath the sun. So did her attacker. Both of them wore the clothes that make for an instantly recognizable uniform: shirts and pants weathered by hard daily wear to a neutral color and texture. Her possessions, and his, were stuffed into two bulky backpacks that leaned against a signpost in a median dividing six lanes of city traffic. I was in the middle lane of traffic on one side. All of the other people in this crowd were also in automobiles, on either side of me, opposite me, ahead and behind, most of them with their windows rolled up, listening to the radio or talking on cell phones. From what I could tell, no one else was watching this woman get beaten up and chased across three, then six, then nine separate lanes of traffic in the intersecting streets. I considered how I could get out of my car (should I leave it idling? lock it? what?) and run toward this woman and man, shouting at him to stop, begging the other drivers to use their phones to call the police. And then, after I had turned over this scenario in my mind for eight or nine seconds, the light changed and every car but mine began to move, and I had to think instead about the honking horns, the blocked traffic, the public nuisance I was about to become, and all the people who would shake their heads at my do-gooder foolishness and inform me that I should stay away from these rough-looking characters because this was obviously a domestic dispute.

But that could not have been true. It was not domestic. *Domestic* means "of the home," and these people had no home. That was the problem—theirs, mine, everybody's. These people were beneath or somehow outside the laws that govern civil behavior between citizens of our country. They were homeless.

In his poem "Death of a Hired Man," Robert Frost captured in just a few words the most perfect definition of home I've ever read:

> *Home is the place where, when you have to go there,*
> *They have to take you in.*

I wish I could ever have been so succinct. I've spent hundreds of pages, even whole novels, trying to explain what home means to me. Sometimes I think it's the only thing I ever write about. Home is place, geography, and psyche; it's a matter of survival and safety, a condition of attachment and self-definition. It's where you learn from your parents and repeat to your children all the stories of what it means to belong to the place and people of your ken. It's a place of safety—and that is one of the most real and pressing issues for those who must live without it. For homeless women and men, the probability of being sexually assaulted or physically attacked is so great that it's a matter not of *if* but of *when*. Homelessness is the loss first of community and finally of the self. It seems fatuous that I could spend so much time contemplating the subtle nuances of home (let alone buy a magazine devoted to home remodeling or decor) when there are people near me—sometimes only a few feet away from me—who don't have one, can't get one, aren't even in the picture.

Tucson, the city where I live most of the time, is often said to be a Mecca for homeless people. I don't like this use of the term "Mecca" because it suggests a beautiful holy land and a trip undertaken to fulfill the needs of the spirit. Whereas thousands of men, women, and children undertake the long journey to Tucson each winter for one reason only: to fulfill the need of the body to lie on the ground overnight without freezing to death. They come here for raw survival; their numbers swell each October and then remain through each following summer a little higher than the year before. Increasingly they have become a presence among us, ignored by most of us, specifically banned by our laws from certain places where the city council has decided their panhandling may interfere with commerce. They are banned mostly, I think, because their presence is a pure, naked shame upon us all.

Whatever else "home" might be called, it must surely be a fundamental human license. In every culture on earth, the right to live in a home is probably the first condition of citizenship and humanity. Homelessness is an aberration. It may happen anywhere from time to time, of course, but when I look hard at the world, I see very few places where there resides an entire, permanent class of people labeled "homeless." Not in the poorest places I've ever lived, not even in an African village where everyone I knew owned only one shirt (at best) and most had never touched an automobile. Because even there, as long as the social structure remains intact, people without resources are taken in by their families. Even if someone should fall completely apart and have to go to the hospital, which means a trek on foot over dozens of miles or more, the whole family goes along to make sure the sick one is taken care of. "Home," in this case, becomes portable. I know this because I lived as a child in an African village that housed the region's small, concrete-block hospital. Whenever I walked past, the hospital's lively grounds never failed to impress me. It was just a bare-dirt plaza, maybe stretching among all its corners to the size of a city block, but it was always a busy place, where dozens of families camped out around their cooking fires while waiting for some relative to have an operation, have a baby, or die. Meanwhile they passed the time by singing, mourning, washing dishes, arguing, daydreaming, or fussing at toddlers who ran around wearing nothing but strings of beads around their bellies. In the rest of my life I have never witnessed another scene so solidly founded on both poverty and security. I don't wish to glorify the impoverished half of this equation; these children had swollen bellies from kwashiorkor, and they had parasites. But they also had families they could not forget under any circumstances, or ever abandon, or be abandoned by, however they might fall on madness or illness or hard times. I don't believe that the word *homeless* as it's used in our language could be translated there.

In most cultures I have known or read about, the provision of home is considered to be the principal function and duty of human family. In rich countries other than ours, such as Japan, the members of the European Community, and Canada, the state assumes this duty as well; their citizens pay higher taxes than we do, and so the well-off live with a little less. Generally speaking, the citizens of these nations have smaller homes, smaller cars, and smaller appetites for consumer goods than we do. And to balance it out they have a kind of security unknown to U.S. citizens—that is, the promise that the state will protect every citizen from disaster. A good education, good health care, and good shelter are fairly well guaranteed, even to those who have devastating illness or bad luck. The Revised European Social Charter of the Council of Europe (1996) states in Article 34: "In order to combat social exclusion and poverty, the Union recognizes and respects the right to social and housing assistance to ensure a decent existence for all those who lack sufficient resources." More recently, at the Lisbon Summit in March 2000, heads of the fifteen European Union countries agreed to develop a common strategy for providing universal access to decent and sanitary housing. These civilized nations have long agreed that homelessness simply isn't an option.

Where *does* homelessness exist? On the border between the Congo and Rwanda when those countries are engaged in protracted civil war. In Kosovo, for the same reason. In India, wherever the construction of a massive dam has inundated villages. In Kenya and other parts of Africa, where large numbers of children have lost their entire extended families to AIDS. Many were also homeless in Somalia during the drought, in the Philippines after the volcanic explosion, in Mexico City after the earthquake. In other words, homelessness as a significant problem occurs in countries stricken by war, famine, plague, and natural disaster. And here, in the USA. Why are we not carrying on with ourselves, our neighbors, and the people who represent us the conversation that begins with the question, *What on earth is wrong with us?*

This is a special country, don't we know it. There are things about the way we organize our society that make it unique on the planet. We believe in liberty, equality, and whatever it is that permits extravagant housing developments to be built around my hometown at the rate of one new opening each week ("Model homes, 6 bedrooms, 3-car garages, starting from the low $180s!"), while fully 20 percent of children on my county's record books live below the poverty level. Nationwide, though the homeless are a difficult population to census, we can be sure they number more than one million. How does the rest of the world keep a straight face when we go riding into it on our latest white horse of Operation-this-or-that-kind-of-Justice, and everyone can see perfectly well how we behave at home? Home is where all justice begins.

More than a decade ago, a government study discovered the surprising fact that some 10 percent of American families appeared to be destined for homelessness. These were working families with a household income, not qualifying for unemployment or other kinds of relief, but they had to spend more than 60 percent of that income on housing and heating; 44 percent of it on food; and 14 percent on medical care. It doesn't take higher math to show they were having to go into debt, deeper each month, to stay alive. This truth was demonstrated dramatically in more recent years by Barbara Ehrenreich, who gave two years of her life at the end of the 1990s to the challenge of surviving on the best wages she could find as an unskilled worker, and then writing about it in her remarkable book *Nickeled and Dimed: On (Not) Getting By in America.* When workers earning minimum wage can no longer meet their bills, they must decide which of the following things to give up: housing, food, medical care. That is one of the several life histories we call, collectively, the American way.

The figures above came from a book by Arthur Blaustein, who was appointed by President Carter in 1977 to chair the National Advisory Council on Economic Opportunity. The council was abolished in 1981 by the Reagan administration, which didn't like the council's findings. But Blaustein published the work anyway, in his book *The American Promise*, which outlines our nation's disastrous approach to dealing with poverty. In his introduction he mentions an interview between Bill Moyers and Robert Penn

Warren, a writer I'm proud to claim as a fellow Kentuckian, who was at that time America's Poet Laureate. Mr. Moyers asked Mr. Warren, "Sir, as one of our leading writers and philosophers, can you tell me how we can resolve the terrible crises that surround us: decaying cities, terrible health care, terrible crises in education and housing, and so much poverty?"

Mr. Warren leaned forward and said, "Well, Bill, for a beginning, I think it would be good if we would stop lying to one another."

This is it. This is all. We so desperately avoid looking at the truth square on, much less saying it aloud, because it's uncomfortable for us to go about our days in relative luxury while people next door to us are dying for lack of shelter. Civic pride can lose its shine when reality is allowed a place at the table. I find it unspeakably hard to walk past someone whose life would be improved, noticeably, by the amount of spare change I could probably find on the floor of my car.

But we manage, those of us who are lucky enough, to walk on by. We live with pronounced class difference in a nation that was founded on the ideal of classlessness, and we do it by believing in a comforting mythology of genesis that is as basic to our nation as the flag and the pledge of allegiance. Here are some of our favorite fuzzy-blanket myths:

1. *Anybody* who is clever and hardworking can make it in America.
2. Homeless people are that way for some good reason. They chose it, or they're criminals or alcoholics or crazy, but whatever went wrong, it's their fault. It couldn't happen to me because I'm clever, sane, and hardworking.
3. Or maybe it isn't *entirely* their fault. But the problem of poverty is so complex that it's impossible to fix.

As a professional storyteller, I take myths personally. I take it as part of my job to examine the stories that hold us together as a society and that we rely on to maintain our identity. These particular myths about poverty are probably some of the most useful tales that create our cultural persona. I also think they're individually destructive and frankly untrue, and oh, yes, they kill people. Finally, this mythology is omnipresent, embedded to some degree in virtually every heart—healthy, wealthy, and otherwise—that beats within this union. Rich people may believe it and relax; poor people may believe it and become paralyzed with self-loathing. And the rest of us just muddle on. When I drive my car through an intersection, past four more homeless men or women (out of the thirty or more I might see in a day) with four more cardboard signs poised to drive a stake of guilt through my heart, I can hear these quiet words rising up in the back of my mind:

". . . smart like me, . . . hardworking like me . . . they'd have a house like me."

And here are some questions I have occasionally had to ask myself, as a counterpoint to that little song: Am I so smart that I could survive on my wits alone, without shelter, for months or years? Could I face the enormity of that loneliness and despair without emotional painkillers in the form of alcohol or drugs?

Doubtful.

Am I *hardworking* enough that I could walk ten or fifteen miles every day in the blazing sun from a shelter downtown or a camp along the riverbank to get to this intersection, and then stand here begging? Could I stand on my own two feet all day on this scorched white pavement without water or food or shade or an ounce of love, through the 105-degree heat of every June, July, and August day in Tucson, Arizona?

No.

And then I try to imagine for just a moment that I am God, or at any rate someone kindhearted and smart who is in a position to look down from above on this scenario: Four men are standing hatless on the four corners of the busy intersection at Speedway and Tucson Boulevard while a hundred passengers eye them with indifference from idling air-conditioned cars, or glance away, waiting for the light to change. Who in this scene is clever? Who is lazy? Which four people worked hardest this day to get where they are right now and to stay alive?

This problem is not complicated. First, it might be useful for us to take the advice of a wise old Kentuckian and stop lying to one another. We live in the only rich country in the world that still tolerates this much poverty in the midst of that much wealth. The European Community members and other industrialized nations have declared themselves unwilling to tolerate homelessness, and they devote the resources necessary to guarantee a "decent existence" for all. We could do the same here without all of us first having to study trigonometry or rewrite the Constitution with our bare hands. It would just take money and a shift in values. Our elected officials could allocate the money to this instead of cutting the taxes of corporations and the wealthy, and they *will*—if and when enough of their constituents demand it. In the meantime it is possible to reallocate some money with civilian hands by writing a check or volunteering. Shelters, which offer enough beds for fewer than half the people who use these services, receive about 65 percent of their funding from federal, state, and local governments but are kept in operation almost entirely by volunteer labor. This means more than kitchen work, because humans don't live by soup alone. Volunteers teach music, literacy, and job skills, plan activities for children, and register homeless people to vote. There is nothing so educational as conversing with someone who has lost the condition of home and finding no hard boundaries of virtue that divide you. As a friend who is wheelchair-bound sometimes reminds me, "Barbara, the main difference between you and me is one bad fall off a rock."

I wish I could go back to that afternoon that haunts me and do what I know I should have done: get out of my car, make a scene, stop traffic, stop a violent man if I could. Home is the place where, when you have to go there, they have to take you in. My car might have been the place she had to go, with no other earthly alternatives left to her, and so it may be that I have to take her in, take that risk, get criticized or tainted by the communicable disease of shame that is homelessness. In some sense she did come in, for she is still with me. I rehearse a different scene in my mind. If I meet her again I hope I can be ready.

It's a tenuous satisfaction that comes from rationalizing problems away or banning them from the sidewalk. Another clean definition I admire, as succinct as Frost's for the complexities of home, is Dr. Martin Luther King Jr.'s explanation of peace: True peace, he said, is not merely the absence of tension. It's the presence of justice.

What Good Is a Story?

I have always wondered why short stories aren't popular in modern America. We are such busy folk, you'd think we'd jump at the chance to have our literary wisdom served in doses that fit between taking the trash to the curb and waiting for the carpool. We should favor the short story and adore the poem. But we don't. Short-story collections rarely sell half as well as novels; they are never blockbusters. They are hardly ever even block-denters. From what I gather, the typical American reader (let's call him

Fred) would sooner plow through a five-hundred-page book about southern France or a boy attending wizard school or how to make home decor from roadside trash or *anything* before he'll pick up a tale of the world complete in twenty pages. And I won't even discuss what Fred will do to avoid reading poetry.

Why should this be? I enjoy the form so much that when I was invited to be the guest editor for a special story collection, forewarned that it would involve reading thousands of pages of short fiction in a tight three-month period, I decided to do it. This trial by fire would disclose to me (I thought) the heart of the form and all its mysteries. Also, it would nicely fill the space that lay ahead of me at the end of the year 2000, just after my intended completion of the novel I was working on and before its scheduled publication the following spring. The creative dead space between galley proofs and a book's first review is a dreaded time in an author's life, comparable to the tenth month of a pregnancy. (I've had two post-term babies, so I know what I'm talking about here.) I regard the prepublication epoch as a Great Sargasso Sea on my calendar and always try to fill it with satisfying short-term projects. A writer who'd edited the same anthology in an earlier year described the organized pleasure of reading one story a day for three months. That sounded like a tidy plan to put on my calendar. Editing a story collection, plus a brief family vacation to Mexico and a week-long lecturing stint on a ship in the Caribbean, would fill those months perfectly, providing just enough distraction from my prepublication doldrums.

If you ever want to know what it sounds like when the universe goes "Ha! Ha!," just put a tidy plan on your calendar.

My months of anticipated quiet at the end of 2000 turned out to be the most eventful of my life, a period in which I was called upon to attend an astounding number of unexpected duties, celebrations, and crises. I weathered a publicity storm with the release of my new novel eight months ahead of schedule. I decided to turn the proposed book tour into a series of fund-raisers for environmental organizations, and had the privilege of working with dedicated advocates of this continent from one threatened coastline to the other. While handling this, plus the lectures at sea, I was invited to receive a national medal and have dinner with President and Mrs. Clinton. Right in the middle of it all, we learned of a family member's catastrophic illness. And then, stunned by still more unexpected grief, I took my eighth grader to the funeral of a beloved friend. This is not even to mention the normal background noise of family urgencies. These two months of our lives were stitched together by trains, automobiles, the M.S. *Ryndam*, and thirty-two separate airplane flights (a perverse impulse caused me to save my boarding passes and count them). Naturally this would be the year when I also experienced a true airplane emergency, and I don't mean the garden-variety altitude plunge. I mean that I finally got to see what those yellow masks look like.

Through it all, as best I could, I read stories. On a cold Iowa afternoon, with the white light of snowfall flooding the windows, sitting quietly with a loved one enduring his new regime of chemotherapy, I read about a nineteenth-century explorer losing his grasp on life in the Himalayas. On another day, when I found myself wide-eyed long after midnight on a ship at sea so racked by storms that the books were diving off the shelves of my cabin, I amused myself with a droll fable about two feuding widows in the Pyrenees. I read my way through a long afternoon sitting on the dirty carpet of Gate B-22 at O'Hare, successfully tuning out all the mayhem and canceled-flight refugees around me except for one young woman who kept shouting into her cell phone, "I'm almost out of minutes!" (This was not the same day my airplane would lose its oxygen; the screenwriter of my life isn't *that* corny.) I read through a Saturday while my four-year-old dozed in my lap with a mysterious fever that plastered her curls to her forehead and burned my skin through her pajamas; I read in the early mornings in Mexico while parrots chattered outside our window. Some days I was able to read no stories at all—when my young daughter was *not* asleep on my lap, for instance—and on other days I read many. Eighteen stories got lost with my luggage and took a trip of their very own, but returned to me in time.

My ideas about what I would gain from this experience collapsed as I began to wrestle instead with what I would be able to give to it. How could I read 125 stories amid all this craziness and compare them fairly? In the beginning I marked each one with a ranking of minus, plus, or double-plus. That lasted for exactly three stories. It soon became clear that what looks like double-plus on an ordinary day can be a whole different story when the oxygen masks are dangling from the overhead compartment. I despaired of my wildly uncontrolled circumstances, thinking constantly, If this were my writing, would I want some editor reading it under these conditions?

Maybe not. But the problem is, life is like that. Editors, readers—all of us have to work reading into our busy lives. The best tales can stand up to the challenge—and if anything can, it should be the genre of short fiction with its economy of language and revving plot-driven engine. We catch our reading on the fly, and that is probably the whole point, anyway. If we lived in silent white rooms with no emergencies beyond the wilting of the single red rose in the vase, we probably wouldn't need fiction to help us explain the inexplicable, the storms at sea and deaths of too-young friends. If we lived in a room like that we would probably just smile and take naps.

What makes writing good? That's easy: the lyrical description, the arresting metaphor, the dialogue that falls so true on the ear it breaks the heart, the plot that winds up exactly where it should. But these stories I was to choose among had been culled from thousands of others, so all were beautifully written. My task was to choose, from among the good, the truly great. How was I supposed to do it? With a pile of stories on my lap I sat with this question, early on, and tried to divine for myself why it was that I loved a piece of fiction when I did, and the answer came to me quite clearly: I love it for what it tells me about life. I love fiction, strangely enough, for how true it is. If it can tell me something I didn't already know, or maybe suspected but never framed quite that way, or never before had sock me so divinely in the solar plexus, that was a story worth the read.

From that moment my task became simple. I relaxed and read for the pleasure of it, and when I finished each story, I wrote a single sentence on the first page underneath the title. Just one sentence of pure truth, if I found it, which generally I did. No bumpy air or fevers or chattering parrots could change this one true thing the story had meant to tell me. That was how I began to see the heart of the form. While nearly all the stories were pleasant to read, they varied enormously in the weight and value of what they carried—whether it was gemstones or sand that I held in my palm when the words had trickled away. Some beautifully written stories gave me truths so self-evident that when I wrote them down, I was embarrassed. "Young love is mostly selfish," some told me, and others were practically lining up to declare, "Alcoholism ruins lives and devastates children!" In the privacy of my reading, I probably made that special face teenagers make when forced to attend to the obvious. Of all the days of my life, these were the ones in which I was perhaps most acutely aware that time is precious. Please, tell me something I don't already know. Sometimes I couldn't find anything at all to write in that little space under the title, but most of the stories were clear enough in their intent, and many were interesting enough to give me pause. And then came one that rang like a bell. I knew this story had given me something I would keep. I slipped it into a pocket of my suitcase, and when I got home I set it on the deep windowsill beside my desk where the sun would fall on it in the morning and over two months it would grow, I hoped, into a pile of stories of equal value. Words that might help me become a better mother, a wiser friend. I felt I'd begun a shrine to new truths, the gifts I was about to receive in a difficult time.

Slowly that pile did grow. Too slowly, I feared at first, for when I'd conquered nearly half my assigned reading, it still seemed very small. I am too picky, I thought. I should relax my standards. But how? You don't "lower the bar" on enlightenment. I couldn't change my heart, so I didn't count the stories in my shrine, I just let them be what they were. Cautiously, though, I made another pile called "Almost, maybe." If push came to shove, I would reread these later and try to be more moved by them.

If it sounds as if I'm a terribly demanding reader, I am. I make no apologies. Long before I ever heard the words, "We're going to try an emergency landing at the nearest airport that can read our black box" (I swear this really happened; that pilot should go to charm school), it had already dawned on me that I wasn't going to live forever. This means I may never get through the list of great books I want to read. Forget about bad ones, or even moderately good ones. With *Middlemarch* and *Pilgrim at Tinker Creek* in the world, a person should squander her reading time on fashionably ironic books about nothing much? *I am almost out of minutes!* I'm patient with most corners of my life, but put a book in my hands and suddenly I remind myself of a harrowing dating-game shark, long in the tooth and looking for love *right now,* thank you, get out of my way if you're just going to waste my time and don't really want kids or the long-term commitment. I give a novel thirty pages and if it's not by that point talking to me of till-death-do-us-part, then sorry, buster, this date's over. I've chucked many half-finished books into the donation box.

You may be thinking right now that you're glad I was never your writing instructor, and a few former students of mine would agree with you. Once in a workshop after I'd repeatedly explained that brevity is the soul of everything, writing-wise, and I was still getting fifty-page stories that should have been twenty-page stories, I announced, "Starting tomorrow I will read twenty-five pages of any story you give me, and then I'll stop. If you think you have the dazzling skill to keep me hanging on for pages twenty-six-plus because my life won't be complete without them, just go ahead and try."

I'm sorry to admit I was such a harpy, but this is a critical lesson for writers. We are nothing if we can't respect our readers. It's audacious, really, to send a new piece of writing out into the world (which already contains *Middlemarch*) asking readers to sit down, shut up, ignore kids or work or whatever important irons they have in the fire and listen instead to *me*. Not just for a minute but for hours, days. Whatever I've got to say had better be important, worth every minute you're giving it, with interest.

Probably the greatest challenge of the short-story form is to get a story launched and landed efficiently with a whole, worthwhile journey in between. The launch is apparently easier than the landing; I've been entranced by many a first paragraph of a tale that ended with such an unfulfilling thud as to send me scrambling around looking for a next page that simply wasn't. Maybe the average American doesn't read short stories simply because of a distaste for this kind of a ride. A good short story cannot be simply Lit Lite. It should pull off the successful execution of large truths delivered in tight spaces. If all short fiction did this perfectly, or even partially, then Fred would surely read more of them.

For me to love a work of fiction, it must survive my harpy eye on all accounts: It will tell me something remarkable, it will be beautifully executed, and it will be nested in truth. The latter I mean literally; I can't abide fiction that fails to get its facts straight. I've tossed aside stories because of botched Spanish or French phrases uttered by putative native speakers who were not supposed to be toddlers or illiterates. More often, I've stopped reading books in which birds sang on the wrong continents or full moons appeared two weeks apart (no, it wasn't set on Jupiter). I am not sure whether the preponderance of scientific howlers in fiction derives from the fact that most writers don't take science courses in school, or if I just notice them more because I *did* take the science courses. In any case people learn from what they read, they trust in words, and this is not a responsibility to take lightly. Scientific illiteracy is a problem I care about, no matter whether it comes from inadequate science instruction or from nonscientists' playing fast and loose with facts. Literature should inform as well as enlighten, and first, do no harm.

I ask a lot from my reading—ask of it, in fact, what I ask of myself when I sit down to write, and that is to get straight down to the task and carve something hugely important into a small enough amulet to fit inside a reader's most sacred psychic pocket. I don't care what it's about, so long as it's not trivial. I once heard a writer declare from a lectern, "I write about the mysteries of the human heart, which is the only thing a fiction writer has any business addressing." And I thought to myself, Excuse me? I had recently begun thinking of myself as a fiction writer and was laboring under the illusion that I could address any mystery that piqued me, including but not limited to the human heart, human risk factors, human rights, and why some people practically have to scrape flesh from their bones to pay the rent while others have it paid for them all their merry days, and how frequently the former are women raising children by themselves, even though that wasn't the original plan. The business of fiction is to probe the tender spots of an imperfect world, which is where I live, write, and read. I want to know about the real price of fast food in China, who's paying it, and why. I want to know what it's like in Chernobyl all these years later. Do you? I've learned the answers from short stories.

When I look back now on the three months that launched a

hundred travails of my heart and in the course of which I read more than a hundred stories, I understand that the reading was really not just a chore piled onto an unbelievably overscheduled piece of my life. Rather, it was a kind of life raft through it. While the people around me in Gate B-22 swore irritably through their cell phones, I was with a man in an Iranian prison who survived isolation by weaving a rug in his mind. The night after my teenager and I returned from her friend's funeral and she asked me how life could be so unfair, I lay down on my bed to read of the pain and healing of a child from Harlem. The stories were, for me, both a distraction and an anchor. Good fictional tales will always be my pleasure, my companionship, my salvation. I hope they're also yours.

Marking a Passage

In 1958, when I was hardly yet a dot on any map, a new bookstore opened its doors at the eastern end of the main street of what would one day become my hometown, Tucson, Arizona. The bookstore filled its shelves, and customers happily bought what was there, then asked for more, until the Book Mark came to assume the higgledy-piggledy atmosphere I associate with old London bookshops: high bookshelves stacked even higher on top with oversize stock lying sideways; sliding ladders shoved up and down

narrow aisles crammed with every kind of wisdom. And always, of course, the friendly staff waiting to help track it down. One of them was a tiny woman named Anne whose memory presaged computers. She could identify just about anything ever published, declare it in stock, and scramble up a ladder for it, hanging up there near the ceiling fixtures and chattering away about what *else* this author had written, while your heart quailed lest she fall and dash her frail bones.

Twenty-five years later, when I was still on nobody's map but my own, I had claimed that marvelous bookstore as my territory: I met friends there, attended my first literary readings, gained a little confidence at debating art and politics, gave and received recommendations for the obscure but spectacular first novel that would have to be read before anyone's life went one step farther. Once in this bookish meeting place I even began a star-crossed love affair behind the discreetly turned spines of Virginia Woolf and Leo Tolstoy.

And then came the year 1988, when, unbelievably (to me, anyway), I was about to publish an obscure first novel of my own. My New York publisher turned out a few thousand copies, and we all hoped most of them would sell before it faded out of print. This is the way of first novels, which aren't generally greeted with trumpets. Mostly they are greeted with yawns. That is why most writers starve, or else have day jobs.

I was lucky, though. I had a guardian angel, a tiny one named Anne, who loved my book and made it her mission to shove it into the hands of everyone she thought would like it. This meant pretty much every unarmed human being who entered the Book Mark, and some who were merely hanging around in the parking lot. I had other guardian angels, too, it turns out: booksellers all over the country who discovered my novel and sold it "by hand," as they say in the business.

Booksellers proceeded to change my life, and they have changed it since then in some ways they've probably never imagined. My name, for instance, used to be a disaster that nobody could spell. You don't just blurt out a name like mine, because it leaves people breathless and staring as if you'd made an inappropriate noise. For about thirty years I never said my name without spelling it: "Kingsolver—K-I-N-G-S as in Sam-O-L-V as in Victor-E-R—yes, ma'am, just like it sounds." That's twenty-six syllables. I still have to do that sometimes, but not at the library, no indeed, and not when I call to place an order at a bookstore. "Kingsolver, like the author?" they'll ask. "Any relation?" And I'll say yes, I think so. If I'm feeling sassy I'll say, "Yes, I'm married to her husband." How can I measure what this means in my life, to have a last name that's been reduced from twenty-six syllables to three?

And that's nothing at all compared with the joy of working at a job I love. I'm finally pretty confident that I have quit the last of the long string of day jobs I held for so many years to support my writing, and now—thirteen years later—I don't think any of my old bosses is expecting me back. Each day as I sit down to work on writing a book, I begin by laying my hands gently across my keyboard and offering up my silent thanks to readers, to the people who publish books, and to the people who sell them—the people without whom I would not get to do what I do.

When my first novel came out in its three-or-four-thousand-copy edition, my publisher and I crossed our fingers for luck, because that's how it is for little first novels with funny names on their spines that nobody can spell. After my guardian angels began to press it into people's hands, it went back to print, then back to print again, and while it didn't break any records for a first novel, it got read, all over the place. I earned enough royalties so I didn't have to go back to my day job to feed my baby and keep up with the mortgage. Instead, I got to stay at my desk and write a second book, then a third and a fourth. I finished each one with the help of booksellers who were rooting for my career. Now the copies of my books out in the world number in many millions, a lot of them in languages I can't read. This strikes me as a miracle on the order of the loaves and fishes.

Ten years after that first precarious launch of a first novel, my friends at the Book Mark asked me to give the debut reading of my eighth book, *The Poisonwood Bible,* at their store, and I told them I couldn't imagine any better place to launch it. I stood on a raised platform in their parking lot, surrounded by hundreds of cheering Tucsonans, and felt a little like Evita Peron. I swore I would never forget that day or the people there who had first guided readers to my words, and to whom I knew, absolutely, I owed my career. I thought of them as family. When my second daughter was born I sent them a birth announcement, which they proudly displayed.

Then, the February following that marvelous coronation in their parking lot, they sent me a much less joyful announcement: After forty years, the Book Mark was passing away. Tucsonans' buying habits were changing, it seemed. They were purchasing through the Internet, hunting for bargains, and drawn by the lure of chain stores.

Over the next weeks I determined to go in often to say good-bye to my favorite aisles and buy more books from their emptying shelves. The store had to sell as much of its stock as it could, I realized, but I dreaded that on my visits I would feel as if I were sifting through the goods of a dying relative. Nevertheless, I made myself march down to my bookstore with the same cheer and courage that my old friend Anne, now deceased, had once brought to the project of hand-selling my first novel. I hugged each of my friends behind the counter and told them: I can't bear this passing.

I couldn't, and still can't, because the scene is repeating itself in cities everywhere as other small and large independents announce their final closing sales. I'm grateful, of course, that books are still sold elsewhere, in other stores including the national chains, and I know that in some small towns that have never before had the privilege of having a real bookstore, the nationally run stores now turning up may be a godsend. I appreciate the reading series and book clubs they have organized where such things have never happened before. The big stores have their place; I'd just be happiest if it weren't *in place of* the other kind. I have a bone to pick with any behemoth if its strategy includes purposefully locating close to, and outcompeting, the neighborhood shops by offering discounts on its most popular stock. I am uncomfortable with taking advantage of a bargain if the store's size allowed it to snag that book for a reduced price from a publisher that didn't offer the same deal to the independents. (Independent booksellers have challenged this practice in court.) Publishers also subsidize certain books by giving "co-op money" to all booksellers; the chains get these subsidies on a scale that determines which books will move forward into the large, front-of-the-store displays. This practice means that people in many cities at once will hardly be able to walk in the door of their nearest big bookstore without tripping over a stack of the new Stephen King or—yes—the latest Barbara Kingsolver.

I am stunned and flattered to find myself so prominently displayed, and deeply grateful for support I've received from any and all bookselling quarters. But I'm humbled by what I know of my roots: I wasn't always up there, front and center. Once I was the name no one could spell, on the spine of a book that could have gone quietly out of print while its author went back to free-lance science writing or professional (but cheerless) housecleaning. There but for the grace of my guardian angels go I. If I were trying to launch a writing career today, I would be launching it into very different waters. I could not possibly have as much support from independent booksellers as I did back then, simply because there are not as many of them now. I'd be taking more of my chances in the chains, as an impossible name on the spine of a little lost book somewhere in the back of the store—a needle in a brightly lit haystack. I can't be sure I'd be a writer today if that's what I was up against, starting out. What I'm sure of is this: Mine would have been much more of an uphill struggle without my legions of bookseller-promoters, and I would not have been able to write as many books. Some of the titles I've given the world would not be in it. There would be no glorious launching of *The Poisonwood Bible* in the parking lot of the Book Mark; there would be no Book Mark, and no *Poisonwood Bible*. I don't like this grim re-visioning of my life as it might have been, but it's the truth.

It's not only starving artists who should care about what we're losing when an independent bookstore dies. This is not about retail; it's about people who serve as community organizers in places where you can always find kindred spirits, a good read, maybe even love behind the spine of Virginia Woolf. A store where you can be sure no one will say to you, as happened to someone I know when he went into a place I shall not name, asking for *Catcher in the Rye*: "Um, check the sports aisle?"

Putting an extra dollar or two back into our hometown's economy, rather than sending it off to a distant, faceless conglomerate, is worth what it costs for so many reasons. Holding on to our independent booksellers is nothing less than a First Amendment issue. To put it bluntly, megasellers and megapublishers have significant power, when put together, to manipulate what Americans will see, purchase, and read. Their power has its purpose, but it needs to be balanced. "Independent" means what it says: stores that are locally owned, by people who know books and need not tailor their orders to the appetites of a distant city, but would rather honor their customers' interests in regional issues, local authors, small-press books, poetry, first novels—things that matter to *us*, right here, right now. Is this something you can live without?

Apparently, the answer for lots of us is that yes, we can, and we will have to. Miraculously, Tucson still has a glorious feminist independent named Antigone Books that's still going strong after more than twenty years, as well as a raft of specialty and used-book stores. But most of the rest have gone the way of Arizona's native fish: One by one their streams dried up, and they went extinct. Wonderful names—the Haunted Bookshop, Coyote's Voice, Marco Polo, Whiz Kids—are now a kind of secret code that passes poignantly between old-time Tucsonans who love to read. And now another has joined them—that bookish trout that swam upstream for so long, the Book Mark.

For those sad last weeks of its closing sale I was stuck in the earliest stage of grief: denial. I kept banking on a miracle on the order of Jimmy Stewart's in *It's a Wonderful Life*. People would show up there in droves with cash in hand, I thought, to prove that their hearts had not been sold after all for the three-dollar markdown. The prodigal readers would return, and those who had never left would also come back to scour the aisles, looking for the enlightenment and passions and how-to manuals that filled our lives before TV stultified and bumfuzzled us. And this actually did happen, in a way: People came in to the store begging to know how they could help, even offering to invest their savings. But for that store, it was too late.

I keep running across the phrase "because of the demise of the independent bookstore," and it causes me to get hot under the collar. The reports of this death are greatly exaggerated—I just can't believe the independents will all go down. The tides of fortune will reverse themselves, I still tell myself, every time I read of another closure. It will happen because this is America, where we love to believe in our own story, the possibility that any one of us could write the Great American Novel, and the rest of us could read it, without waiting for Big Brother to buy it a place at the table. It will happen because we're devoted, above all, to independence and freedom of thought.

Aren't we?

Taming the Beast with Two Backs

Reader, hear my confession: I have written an unchaste novel. It's a little shocking, even to me. In my previous books I mostly wrote about sex by means of the spacebreak. One reviewer claimed I'd written the shortest sex scene in the English language. I know the scene he meant; the action turns when one character notices a cellophane crackle in the other's shirt pocket and declares that if he has a condom in there, this is her lucky day. The scene then proceeds, in its entirety:

He did. It was. [Spacebreak!]

I think my readers have always relied on me for a certain reserve, judging from the college course adoptions and mothers who've said they shared my books with their daughters. They may be in for a surprise this time around. Not that the sex is *gratuitous,* I keep telling myself. This novel is about life, in a biological sense: the rules that connect, divide, and govern living species, including their tireless compunction to reproduce themselves. In this tale the birds do it, the mushrooms do it, and the people do it—starting on page six, already. I found myself having a good old time writing about it, too. I've always felt I was getting away with something marginally legal, inventing fantasies for a living. But now it seems an outright scandal. I send my kids off to school in the morning, scuttle to my office, close the door, and hoo boy, *les bons temps roulent!*

As I closed in on a finished draft, though, I began to think about the people who'd soon be sitting in their homes, libraries, and in subways with their hands on this book. Many people. My mother, for instance.

My writer friend Nancy, a practical New Englander, offered this counsel: "Barbara, you're in your forties now, and you have two children. She *knows* that you *know.*"

Yes, all right, she does. But what about the man from the Ag Extension Service, whom I'd asked to vet my book's agricultural setting for accuracy? How would I hand this manuscript over to him? And what about those English lit teachers? I don't mind that they know I *know,* or that I think about it, in circumstances outside my own experience. Come on, who doesn't? Most people I know couldn't construct a good plot to save their souls—but all of them can and do, I suspect, imagine detailed sexual scenarios complete with dialogue (if they're female) and a sense of place.

But they don't pass them around for others to read, for heaven's sake. My dread is that people will take my book for something other than literature, and me for something other than a serious writer. In my more anxious moments I have combed my bookshelves for the comfort of finding fellow offenders. Yes, there were plenty of authors before me who put explicitly sexual scenes into literature. There's a particularly lovely one in the center of David Guterson's *Snow Falling on Cedars,* there are sweetly funny ones in John Irving, and of course we have John Updike, Philip Roth, and Henry Miller (notice the dearth of females on this list). Even such distinguished eighteenth-century gents as Ben Franklin and Jonathan Swift scored the occasional love scene in their prose. But I was surprised, on the humid afternoon I spent pulling down books and looking for scenes that had burned themselves into my memory, to see how often they were implied situations rather than blow-by-blow enactments. Copious use of the spacebreak, in other words. The scene in *Lady Chatterley's Lover* I'd remembered down the years, it turns out, was invented mostly by me, not by D. H. Lawrence (and given Lawrence's knowledge of love from the female perspective, is that any wonder?). In actual word count, if the literary novels in my bookcase accurately represent human experience, it looks like people spend roughly half their time in intelligent dialogue about the meaning of their lives, and one percent of it practicing or contemplating coition.

Excuse me, but I don't think so.

Why should literary authors shy away from something so important? Nobody else does. If we calibrated human experience on the basis of TV, magazine covers, and billboards, we'd have to conclude that humans devote more time to copulation than to sleeping, eating, and accessorizing the hot new summer look, combined. (Possibly even more than shooting one another with firearms, though that's a tough call.) Filmmakers don't risk being taken less seriously for including sexual content; in fact, they may risk it if they don't. But serious literature seems to be looking the other way, ready to take on anything else, with impunity. Myself, I've written about every awful thing from the death of a child to the morality of political assassination, and I've never felt faint-hearted before. What is it about describing acts of love that makes me go pale? There is, of course, the claim that women who make a public show of being acquainted with sexuality are expressing deviance—but that's also said about women who make a show of knowing *anything,* and I can't imagine being daunted by such nonsense.

For decent folk of any gender, the official and legal position of our culture is that sex takes place in private, and that's surely part of the problem. Private things—newfound love, family disagreements, and spiritual faith, to name a few—can quickly become banal or irritating when moved into the public arena. But new love, family squabbles, and spirituality are rich ground for literature when they're handled with care. We writers don't avoid them on grounds of privacy; rather, we take it as our duty to draw insights from personal matters and render them universal. Nothing could be more secret, after all, than the inside of another person's mind—and that is just where a novel takes us, usually from page one. No subject is too private for good fiction if it can be made beautiful and enlightening.

That may be the rub, right there. Making it beautiful is no small trick. The language of coition has been stolen—or really, I think, it's been divvied up like chips in a poker game among the sides of pornography, consumerism, and the medical profession. None of these players is concerned with aesthetics, so the linguistic chips have become unpretty by association. *Vagina* is thus fatally paired with *speculum.* Any word you can name for the male sex organ or its, um, movement seems to be the property of Larry Flynt. Even a perfectly serviceable word like *nut,* when uttered by an adult, causes paroxysms in sixth-grade boys. My word-processing program's thesaurus has washed its hands of the matter altogether, eschewing any word even remotely associated with making love. "Coitus," for example, claims to be NOT FOUND, with the software coyly suggesting as the nearest alternative: "coincide with?" It also pleads ignorant on "penis," for which it ventures "pen friend?" A writer in work-avoidance mode could amuse herself all day.

I realize that linguistic aesthetics may not be Microsoft's concern here; more likely it's the matter of college course adoptions and mothers. *Roget's* does much better, reinforcing my conviction that the book is mightier than the computer, or at least braver. My St. Martin's *Roget's Thesaurus* obligingly offers up (without even including my favorite from Shakespeare, "the beast with two backs") a whopping *fifteen* synonyms for copulation—I'll admit some of these are dubious, such as *couplement*—and an impressive twenty-eight descriptors for genitalia, though again some of these are obscure. In a scene where lingam meets yoni, I'm not even sure who I'm rooting for.

Nevertheless, the language is ours for the taking. Fiction writers have found elegant ways to describe life on other planets, or in a rabbit warren or an elephant tribe, inventing the language they needed to navigate passages previously uncharted by our tongue. We don't normally call off the game on account of linguistic handicaps. When it comes to the couplement of yoni, I think the real handicap is a cultural one. We live in a strange land where marketers can display teenage models in the receptive lordotic posture (look it up) to sell jeans or liquor, but the basics of human procreation can't be discussed in a middle school science class without risking parental ire. This is also true of evolution, incidentally, and for the same reason, I think: Our religious and cultural heritage is to deny, for all we're worth, that we're in any way connected with the rest of life on earth. We don't come from it, we're not part of it; we *own* it and were put down here to run the place. It's deeply threatening to our ideology, at the corporate and theological levels, to admit that we're constrained by the laws of biology. And yet there it is: sex, the ultimate animal necessity, writhing before us like some alien invader to mission control. We can't get rid of it. The harder we try to deny it official status, the more it asserts itself in banal, embarrassing ways.

And so here we are, modern Americans, with our heads soaked in frank sexual imagery and our feet planted in our puritanical heritage, and any novelist with something to say about procreation or the lordotic posture has to navigate those straits. Great sex is rarer in art than in life because it's harder to do. To broach the subject of sex at all, writers must first face down the polite pretense that it doesn't really matter to us, and acknowledge that in the grand scheme of things, few things could matter more. In the quiet of our writing rooms we have to corral the beast with two backs and find a way to tell of its terror and beauty. We must own up to its gravity. We also must accept an uncomfortable intimacy with our readers in the admission that, yes, we've both done this. We must warn our mothers before the book comes out. We must accept the economic reality that this one won't make the core English lit curriculum.

Still, in spite of everything, I'm determined to write about the biological exigencies of human life, and where can I start the journey if not through this mined harbor? It's a risk I'll have to take.

Reader, don't blush. I know you know.

Stealing Apples

I have never yet been able to say out loud that I am a *poet*.

It took me many years and several published novels to begin calling myself a *novelist,* but finally now I can do that, I own up to it, and will say so in capital letters on any document requiring me to identify myself with an honest living. "Novelist," I'll write gleefully, chortling to think that the business of making up stories can be called an honest living, but there you are. It's how I keep shoes on my kids and a roof above us. I sit down at my desk every day and make novels happen: I design them, construct them,

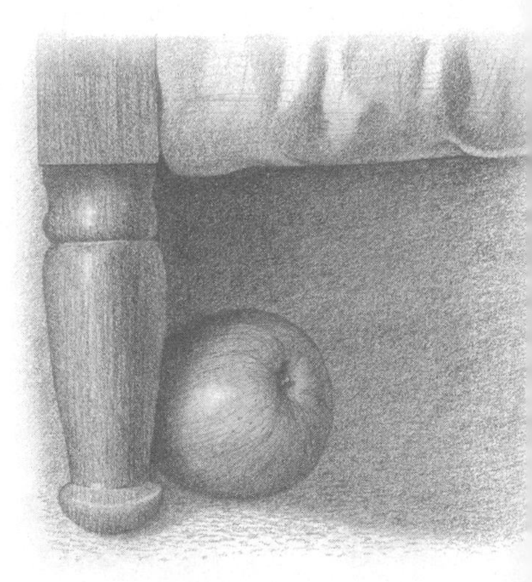

revise them, I tinker and bang away with the confidence of an experienced mechanic, knowing that my patience and effort will get this troubled engine overhauled, and this baby will hum.

Poetry is a different beast. I rarely think of poetry as something I make happen; it is more accurate to say that it happens to *me*. Like a summer storm, a house afire, or the coincidence of both on the same day. Like a car wreck, only with more illuminating results. I've overheard poems, virtually complete, in elevators or restaurants where I was minding my own business. (A writer's occupational hazard: I think of eavesdropping as minding my business.) When a poem does arrive, I gasp as if an apple had fallen into my hand, and give thanks for the luck involved. Poems are everywhere, but easy to miss. I know I might very well stand under that tree all day, whistling, looking off to the side, waiting for a red delicious poem to fall so I could own it forever. But like as not, it wouldn't. Instead it will fall right while I'm in the middle of changing the baby, or breaking up a rodeo event involving my children and the dog, or wiping my teary eyes while I'm chopping onions and listening to the news; *then* that apple will land with a thud and roll under the bed with the dust bunnies and lie there forgotten and lost for all time. There are dusty, lost poems all over my house, I assure you. In yours, too, I'd be willing to bet. Years ago I got some inkling of this when I attended a reading by one of my favorite poets, Lucille Clifton. A student asked her about the brevity of her poems (thinking, I suspect, that the answer would involve terms such as "literary retrenchment" and "parsimony"). Ms. Clifton replied simply that she had six children and could hold only about twenty lines in memory until the end of the day. I felt such relief, to know that this great poet was bound by ordinary life, like me.

I've learned since then that most great poets are more like me, and more like you, than not. They raise children and chop onions, they suffer and rejoice, they feel blessed by any poem they can still remember at the end of the day. They may be more confident about tinkering with the engine, but they'll generally allow that there's magic involved, and that the main thing is to pay attention. I have several friends who are poets of great renown, to whom I've confessed that creating a poem is a process I can't really understand or control. Every one of them, on hearing this, looked off to the side and whispered, "Me either!"

We're reluctant to claim ownership of this mystery. In addition, we live in a culture that doesn't put much stock in it. Elsewhere in the world—say in Poland or Nicaragua—people elect their poets to public office, or at the very least pay them a stipend to produce poetry, regularly and well, for the public good. Here we have no such class of person. Here a poet may be prolific and magnificently skilled, but even so, it's not the *poetry* that's going to keep shoes on the kids and a roof overhead. I don't know of a single American poet who ever made a living solely by writing poetry. That's sad, but it's true. Identifying your livelihood as "Poet" on an official form is the kind of daring gesture that will make your bank's mortgage officer laugh very hard all the way into the manager's office and back. So poets, of necessity, tend to demur. At the most we might confess, "I write poetry sometimes."

And so we do. Whether anyone pays us or respects us or calls us a poet or not, just about any person alive will feel a tickle behind the left ear when we catch ourselves saying, "It was a little big and pretty ugly, but it's coming along shortly. . . ." We stop in our tracks when a child pointing to the sunset cries that the day is bleeding and is going to die. Poetry approaches, pauses, then skirts around us like a cat. I sense its presence in my house when I am chopping onions and crying but not really *crying* while I listen to the lilting radio newsman promise, "Up *next:* The city's oldest homeless shelter shut down by neighborhood protest, *and*, Thousands offer to adopt baby Jasmine abandoned in Disneyland!" There is some secret grief here I need to declare, and my fingers itch for a pencil. But then the advertisement blares that I should expect the unexpected, while my elder child announces that a shelter can't be homeless, and onions make her eyes run away with her nose, and my toddler marches in a circle shouting "Apple-Dapple! Come-Thumb-Drum!" and poems roll under the furniture, left and right. I've lost so many I can't count them. I do understand that they fall when I'm least able to pay attention because poems fall not from a tree, really, but from the richly pollinated boughs of an ordinary life, buzzing, as lives do, with clamor and glory. They are easy to miss but everywhere: poetry just *is,* whether we revere it or try to put it in prison. It is elementary grace, communicated from one soul to another. It reassures us of what we know and socks us in the gut with what we don't, it sings us awake, it's irresistible, it's congenital.

Over the years I've forgotten enough poems to fill several books and remembered enough to fill just one. By the grace of a small, devoted press and a small, devoted contingent of North Americans who read poetry, it remains in print. I began writing poems when I was very young; the most noticeable virtue of my early works was that they rhymed. Then, in high school, I abandoned rhyme scheme in favor of free verse and produced rafts of poems whose most noticeable characteristics were that they were earnest and frequently whiny. I returned again to rhyme scheme and more rigorous structure when I was in college, after seeking my first writing advice from an English professor who advised, "Write sonnets. It will teach you discipline." I dutifully wrote a hundred dreadful sonnets and just one that seemed successful, insofar as its subject suited the extremely confining sonnet form.

Although I had been working at poems and stories all my life, I didn't really begin to understand what it meant to be a writer until my adulthood commenced in Tucson, Arizona, following my arrival here at the very end of the 1970s. I had come to the Southwest expecting cactus, wide-open spaces, and adventure. I found, instead, another whole America. This other America didn't appear on picture postcards, nor did it resemble anything I had previously supposed to be American culture. Arizona was cactus, all right, and purple mountain majesties, but this desert that burned with raw beauty had a great fence built across it, attempting to divide north from south. I'd stumbled upon a borderland where people perished of heat by day and of cold hostility by night.

This is where poetry and adulthood commenced for me, as I understand both those things, because of remarkable events that fell into my quite ordinary way. Oh, I suffered the extremes of love and loss, poverty and menial jobs and exhilarating recuperations, obsessive explorations of a new landscape—all common preoccupations for a young adult in the America I knew. But I also met people, some of them very uncommon. In particular, some of them were organizing the Sanctuary movement, an undertaking I could not previously have imagined in the America I knew. This was an underground railroad run by a few North Americans who placed conscience above law. Their risk was to provide safety for Latin American refugees—many hundreds of them—who faced death in their own countries but could not, though innocent of any crime or ill will, gain legal entry into ours.

I learned, slowly, with horror, that the persecution these refugees were fleeing was partly my responsibility. The dictators of El Salvador, Guatemala, and Chile received hearty support from my government; their brutish armies were supplied and trained by my government. Some of the police who tortured protesters in those countries had been trained in that skill at a camp in Fort Benning, Georgia. My taxes had helped pay for that, and also the barbed wire and bullets that prevented war-weary families from finding refuge here. I wasn't prepared for the knowledge of what one nation might do to another. But that knowledge arrived regardless. I saw that every American proverb had two sides, could be told in two languages, and that injustice did not disappear when I looked away, but instead seeped in at the back of my neck to poison my heart's desires. I saw that unspeakable things could be survived, and that sometimes there was even joy on the other side. I learned all this, one story at a time, from people who had lived enough to know it. Some of them became my friends. Others vanished again, into places I can't know. This great apple that had fallen into my lap became my first novel, *The Bean Trees*. I doubted whether my compatriot readers would really want to hear all that much about what one country would do to another, particularly when one of the countries was ours. I have been wrong about that: once, twice, always.

I believe there are wars in every part of every continent, and a world of clamor and glory in every life. Mine is right here, where I raise my voice and my children, and where we must find our peace, if there is any to be had. Heartbreak and love and poetry abound. We live in a place where north meets south and many people are running for their lives, while many others rest easy with the embarrassments of privilege. Others still are trying to find a place in between, a place of honest living where they can abide themselves and one another without howling in the darkness. My way of finding a place in this world is to write one. This work is less about making a living, really, than about finding a way to be alive. "Poet" is too much of a title for something so incorrigible, and so I may never call myself by that name. But when I want to howl and cry and laugh all at once, I'll raise up a poem against the darkness, or an essay, or a tale. That is my testament to the two boldly different faces of America and the places I've found, or made or dreamed, in between.

One afternoon, as my one-year-old stood on a chair reciting the poems she seems to have brought with her onto this planet, I heard on the news that our state board of education was dropping the poetry requirement from our schools. The secretary of education explained that it took too much time to teach children poetry, when they were harder pressed than ever to master the essentials of the curriculum. He said we had to take a good, hard look at what was essential, and what was superfluous.

"Superfluous," I said to the radio.

"Math path boo!" said my child, undaunted by her new outlaw status.

This one was not going to get away. I threw down my dishtowel, swept the baby off her podium, and carried her under my arm as we stalked off to find a pencil. In my opinion, when you find yourself laughing and crying both at once, that is the time to write a poem. Maybe that's the only honest living there is.

And Our Flag Was Still There

My daughter came home from kindergarten and announced, "Tomorrow we all have to wear red, white, and blue."

"Why?" I asked, trying not to sound anxious.

"For all the people that died when the airplanes hit the buildings."

I said quietly, "Why not wear black, then? Why the colors of the flag, what does that mean?"

"It means we're a country. Just all people together."

I love my country dearly. Not long after the September 11 attacks, as I stood in a high school cafeteria listening to my older daughter and a hundred other teenagers in the orchestra play "Stars and Stripes Forever" on their earnest, vibrating strings, I burst into tears of simultaneous pride and grief. I love what we will do for one another in the name of inclusion and kindness. So I long to feel comforted and thrilled by the sight of Old Glory, as so many others seem to feel when our country plunges into war or dire straits. Symbols are many things to many people. In those raw months following the September 11 attacks, I saw my flag waved over used car and truck lots, designer-label clothing sales, and the funerals of genuine heroes. In my lifetime I have seen it waved over the sound of saber-rattling too many times for my comfort. When I heard about this kindergarten red-white-and-blue plan, my first impulse was to dread that my sweet child was being dragged to the newly patriotic cause of wreaking death in the wake of death. Nevertheless, any symbol conceived in liberty deserves the benefit of the doubt. We sent her to school in its colors because it felt to my daughter like some small thing she could do to help the people who were hurting. And because my wise husband put a hand on my arm and said, "You can't let hateful people steal the flag from us."

He didn't mean foreign terrorists, he meant certain Americans. Like the man in a city near us who went on a rampage, crying "I'm an American" as he shot at foreign-born neighbors, killing a gentle Sikh man in a turban and terrifying every brown-skinned person I know. Or the talk-radio hosts who viciously bullied members of Congress and anyone else for showing sensible skepticism during the mad rush toward war. After Representative Barbara Lee cast the House's only vote against handing over virtually unlimited war powers to a man whom fully half of us—let's be honest—didn't support a year before, so many red-blooded Americans threatened to kill her that she had to be assigned additional bodyguards.

While the anthrax threats in congressional and media offices were minute-by-minute breaking news, the letters of pseudo-patriotism carrying equally deadly threats to many other citizens did not get coverage. Hate radio reaches thousands of avid listeners, and fear stalked many families in the autumn and winter of our nation's discontent, when belonging to *any* minority—including the one arguing for peaceful and diplomatic solutions to violence—was enough to put one at risk. When fear rules the day, many minds are weak enough to crack the world into nothing but "me" and "evildoers," and as long as we're proudly killing unlike minds over there, they feel emboldened to do the same over here. For minds like that, the great attraction to patriotism is, as Aldous Huxley wrote, that "it fulfills our worst wishes. In the person of our nation we are able, vicariously, to bully and cheat. Bully and cheat, what's more, with a feeling that we are profoundly virtuous."

Such cowards have surely never arrived at a majority in this country, though their power has taken the helm in such dark moments as the McCarthy persecutions and the Japanese American internments. At such times, patriotism falls to whoever claims it loudest, and the rest of us are left struggling to find a definition in a clamor of reaction. In the days and months following September 11, some bully-patriots claiming to own my flag promoted a brand of nationalism that threatened freedom of speech and religion with death, as witnessed by the Sikhs and Muslims in my own community, and U.S. Representative Barbara Lee in hers. (Several of her colleagues confessed they wanted to vote the same way she did, but were frightened by the obvious threat from vigilante patriots.) Such men were infuriated by thoughtful hesitation, constructive criticism of our leaders, and pleas for peace. They ridiculed and despised people of foreign birth (one of our congressmen actually used the hideous term "rag heads") who've spent years becoming part of our culture and contributing their labor and talents to our economy. In one stunning statement uttered by a fundamentalist religious leader, this brand of patriotism specifically blamed homosexuals, feminists, and the American

Civil Liberties Union for the horrors of September 11. In other words, these hoodlum-Americans were asking me to believe that their flag stood for intimidation, censorship, violence, bigotry, sexism, homophobia, and shoving the Constitution through a paper shredder? Well, *our* flag does not, and I'm determined that it never will. Outsiders can destroy airplanes and buildings, but only we the people have the power to demolish our own ideals.

It's a fact of our culture that the loudest mouths get the most airplay, and the loudmouths are saying that in times of crisis it's treasonous to question our leaders. Nonsense. That kind of thinking allowed the seeds of a dangerous racism to grow into fascism during the international economic crisis of the 1930s. It is precisely in critical times that our leaders need *most* to be influenced by the moderating force of dissent. That is the basis of democracy, especially when national choices are difficult and carry grave consequences. The flag was never meant to be a stand-in for information and good judgment.

In the wake of the September 11 attacks, an amazing windfall befell our local flag-and-map store, which had heretofore been one of the sleepiest little independent businesses in the city. Suddenly it was swamped with unprecedented hordes of customers who came in to buy not maps, of course, but flags. After the stock quickly sold out, a cashier reported that customers came near to rioting as they stomped around empty-handed and the waiting list swelled to six hundred names. She said a few customers demanded to know why she personally wasn't in the back room sewing more Old Glories. Had I been in her position, I might have said, "Hey, friends and countrymen, wouldn't this be a great time to buy yourselves a map?" The sturdiest form of national pride is educated about the alternatives. And in fairness to my more polite compatriots, I was greatly heartened in that same season to see the country's best-seller lists suddenly swollen with books about Islam and relevant political history.

We're a much nobler country than our narrowest minds and loudest mouths suggest. I believe it is *my* patriotic duty to recapture my flag from the men who wave it in the name of jingoism and censorship. This is difficult, for many reasons. To begin with, when we civil libertarians on the one hand insist that every voice in the political spectrum must be heard, and the hard right on the other hand insists that our side should stuff a sock in it, the deck is stacked. And the next challenge is, I can never hope to match their nationalistic righteousness. The last time I looked at a flag with an unambiguous thrill, I was thirteen. Right after that, Vietnam began teaching me lessons in ambiguity, and the lessons have kept coming. I've learned of things my government has done to the world that make me shudder: Covert assassinations of democratically elected leaders in Chile and the Congo; support of brutal dictators in dozens of nations because they smiled on our economic interests; training of torturers in a military camp in Georgia; secret support even of the rising Taliban in Afghanistan, until that business partnership came to a nasty end. In history books and numbers of our *Congressional Record* I've discovered many secrets that made me ashamed of how my country's proud ideology sometimes places last, after money for the win and power for the show. And yet, when I've dared to speak up about these skeletons in our closet, I've been further alienated from my flag by people who waved it at me, declaring I should love it or leave it. I always wonder, What makes them think that's their flag and not mine? Why are *they* the good Americans, and not me? I have never shrunk from sacrifice but have always faced it head on when I needed to, in order to defend the American ideals of freedom and human kindness.

I've been told the pacifists should get down on their knees and thank the men who gave their lives for our freedom, and I've thought about this, a lot. I believe absolutely that the American Revolution and the Civil War were ideological confrontations; if I had been born to a different time and gender with my present character otherwise intact, I might well have joined them, at least as a medic, or something. (Where I grew up, I'd likely have been conscripted into dying for the wrong side in the Civil War, but that's another story.) I wish I could claim to possess a nature I could honestly call pacifist, but I've had long friendships with genuine pacifists in the Quaker community and have seen in them a quality I lack. I can rarely summon the strength to pray for my enemies, as some do every day. On the rare occasions when my life has been put directly at risk by another, I've clawed like a lioness. My gut, if not my head, is a devotee of self-defense.

But my head is unconvinced by the sleight of hand and sloganeering that put the label "self-defense" on certain campaigns waged far from my bedroom window, against people who have no wish to come anywhere near it. It's extremely important to note that in my lifetime our multitude of wars in Central America and the Middle East have been not so much about the freedom of humans as about the freedom of financial markets. My spiritual faith does not allow me to accept equivalence of these two values; I wonder that anyone's does.

Our entry into wars most resembling self-defense, World War II and the 2001 Afghanistan campaign, both followed direct attacks on our country. The latter, at least, remains a far more convoluted entanglement than the headlines ever suggested. In the 1990s, most of us have now learned, the United States tacitly supported the viciously sexist, violent Taliban warlords—only to then bomb them out of power in 2001. I'm profoundly relieved to see any such violent men removed from command, of course. But I'm deeply uncomfortable, also, with the notion that two wrongs add up to one right, and I'm worried about the next turn of that logic. It is only prudent to ask questions, and only reasonable to discuss alternate, less violent ways to promote the general welfare. Americans who read and think have frequently seen how the much-touted "national interest" can differ drastically from their own.

And Americans who read and think are patriots of the first order—the kind who know enough to roll their eyes whenever anyone tries to claim sole custody of the flag and wield it as a blunt instrument. There are as many ways to love America as there are Americans, and our country needs us all. The rights and liberties described in our Constitution are guaranteed not just to those citizens who have the most money and power, but also to those who have the least, and yet it has taken hard struggle through every year of our history to hold our nation to that promise. Dissidents innocent of any crime greater than a belief in fair treatment of our poorest and ill-treated citizens have died right here on American soil for our freedom, as tragically as any soldier in any war: Karen Silkwood, Medgar Evars, Malcolm X, Denise McNair, Cynthia Wesley, Carole Robertson, Addie Mae Collins, Martin Luther King Jr., Albert Parsons, August Spies, Adolph Fisher, George Engel, Joe Hill, Nicola Sacco, Bartolomeo Vanzetti—the list of names stretches on endlessly and makes me tremble with gratitude. Any of us who steps up to the platform of American protest is standing on bloodstained and hallowed ground, and let no one ever dare call it un-American or uncourageous. While we peace lovers are down on our knees with gratitude, as requested, the warriors might do well to get down here with us and give thanks for Dr. King and Gandhi and a thousand other peacemakers who gave their lives to help lift humanity out of the trough of bare-toothed carnage. Where in the Bill of Rights is it written that the entitlement to bear arms—and use them—trumps any aspiration to peaceful solutions? I search my soul and find I cannot rejoice over killing, but that does not make me any less a citizen. When I look at the flag, why must I see it backlit with the rockets' red glare?

The first time I thought of it that way, I stumbled on a huge revelation. *This* is why the war supporters so easily gain the upper hand in the patriot game: Our nation was established with a fight for independence, so our iconography grew out of war. Anyone who is tempted to dismiss art as useless in matters of politics must agree that art is supremely powerful here, in connecting patriotism with war. Our national anthem celebrates it; our nationalist imagery memorializes it; our most familiar poetry of patriotism is inseparable from a battle cry. Our every military campaign is still launched with phrases about men dying for the freedoms we hold dear, even when this is impossible to square with reality. During the Gulf War I heard plenty of words about freedom's defense as our military rushed to the aid of Kuwait, a monarchy in which women enjoyed approximately the same rights as a nineteenth-century American slave. The values we fought for there are best understood by oil companies and the royalty of Saudi Arabia—the ones who asked us to do this work on the Iraq-Kuwaiti border, and with whom we remain friendly. (Not incidentally, we have never confronted the Saudis about women-hating Wahhabism and vast, unending support for schools of anti-American wrath.) After a swift and celebrated U.S. victory, a nation of Iraqi civilians was left with its hospitals, its water-delivery lines, and its food-production systems devastated, its capacity for reconstruction crushed by our ongoing economic sanctions, and its fate—at the time of this writing—still in the hands of one of the vilest dictators I've ever read about. There's the reality of war for you: Freedom often *loses*.

Stating these realities is not so poetic, granted, but it is absolutely a form of patriotism. Questioning our government's actions does not violate the principles of liberty, equality, and freedom of speech; it exercises them, and by exercise we grow stronger. I have read enough of Thomas Jefferson to feel sure he would back me up on this. Our founding fathers, those vocal critics of imperialism, were among the world's first leaders to understand that to a democratic people, freedom of speech and belief are not just nice luxuries, they're as necessary as breathing. The authors of our Constitution knew, from experience with King George and company, that governments don't remain benevolent to the interests of all, including their less powerful members, without constant vigilance and reasoned criticism. And so the founding fathers guaranteed the right of reasoned criticism in our citizenship contract—for *always*. No emergency shutdowns allowed. However desperate things may get, there are to be no historical moments when beliefs can be abridged, vegetarians required to praise meat, Christians forced to pray as Muslims, or vice versa. Angry critics have said to me in stressful periods, "Don't you understand it's *wartime?*" As if this were just such a historical moment of emergency shutdown. Yes, we all know it's wartime. It's easy to speak up for peace in peacetime—anybody can do that. Now is when it gets hard. But our flag is not just a logo for wars; it's the flag of American pacifists, too. It's the flag of all of us who love our country enough to do the hard work of living up to its highest ideals.

I have two American flags. Both were gifts. One was handmade out of colored paper by my younger child; it's a few stars shy of regulation but nonetheless cherished. Each has its place in my home, so I can look up from time to time and remember, That's *mine*. Maybe this is hard for some men to understand, but that emblem wasn't handed to me by soldiers on foreign soil; it wasn't *handed* to me by men at all—they withheld it from women for our nation's first century and a half. I would never have gained it if everyone's idea of patriotism had been simply to go along with the status quo. That flag protects and represents me only because of Ida B. Wells, Lucy Stone, Susan B. Anthony, and countless other women who risked everything so I could be a full citizen. Each of us who is female, or nonwhite, or without land, would have been guaranteed in 1776 the same voting rights as a horse. We owe a precious debt to courageous Americans before us who risked threats and public ridicule for an unpopular cause: ours. Now that flag is mine to carry on, promising me that I may, and that I must, continue believing in the dignity and sanctity of life, and stating that position in a public forum.

And so I would like to stand up for my flag and wave it over a few things I believe in, including but not limited to the protection of dissenting points of view. After 225 years, I vote to retire the rockets' red glare and the bloody bandage as obsolete symbols of Old Glory. We desperately need a new iconography of patriotism. I propose that we rip strips of cloth from the uniforms of the unbelievably courageous firefighters who rescued the injured and panic-stricken from the World Trade Center on September 11, 2001, and remained at their posts until the buildings collapsed on them. Praise the red glare of candles held up in vigils everywhere as peace-loving people pray for the bereaved and plead for compassionate resolutions. Honor the blood donated to the Red Cross; respect the stars of all kinds who have used their influence to raise funds for humanitarian assistance; glory in the generous hands of schoolchildren collecting pennies, teddy bears, and anything else they think might help the kids who've lost their moms and dads. Let me sing praise to the ballot box and the jury box, and to the unyielding protest marches of my foremothers who fought for those rights so I could be fully human under our Constitution. What could be a more honorable symbol of American freedom than the suffragist's banner, the striker's picket, the abolitionist's drinking gourd, the placards of humane protest from every decade of our forward-marching history? Let me propose aloud that the dove is at least as honorable a creature as the carnivorous eagle. And give me liberty, now, with signs of life.

Shortly after the September attacks, my town became famous for a simple gesture in which some eight thousand people wearing red, white, or blue T-shirts assembled themselves in the shape of a flag on a baseball field and had their photograph taken from above. That picture soon began to turn up everywhere, but we saw it first on our newspaper's front page. Our family stood in silence for a minute looking at that stunningly beautiful photograph of a human flag, trying to know what to make of it. Then my teenager, who has a quick mind for numbers and a sensitive heart, did an interesting thing. She laid her hand over part of the picture, leaving visible more or less five thousand people, and said, "In New York, that many might be dead." We stared at what that looked like—that many innocent souls, particolored and packed into a conjoined destiny—and shuddered at the one simple truth behind all the noise, which was that so many beloved, fragile lives were suddenly gone from us. That is my flag, and that's what it means: We're all just people, together.

God's Wife's Measuring Spoons

Once, not so very long ago, but before I knew how to handle these situations, a reporter came to visit us from the big city for the apparent purpose of finding out what made me tick and revealing it to others. From the start I suspected that this whole thing was not the best idea. I thought about how Georgia O'Keeffe had dealt with a reporter who showed up at her door declaring, "I've come all the way from New York just to see you." Ms. O'Keeffe stood glaring a moment from her doorway and then said, "This is the front," then turned around and said, "This is the back. Now you've seen me." And slammed the

door. I actually tried my own genteel version of that, explaining that I honestly was not all that interesting and she'd be better off interviewing a movie star or something, but this gal was not to be headed off at the pass. Into my life she marched, in sunglasses and snappy shoes, wondering what I was all about.

How would *you* show a person how you tick? I considered giving her a tour of my office, but my writing desk looked the way it usually does: as if a valiant struggle involving lots and lots of papers had recently been fought and lost in there. This theme tends to repeat itself throughout our house—hmm, next the valiant struggle appears to have torn across all the beds, leaving the sheets tangled, then it must have passed through the playroom, touching off forceful eruptions of doll clothing and Legos, before finally exiting out the front door. One end of our dining table looks as if someone's running a mail-order business from it, but I swear it isn't me. Our house reveals about us the same thing my friends' homes do about them: here lives a busy family, most of whom have better things to do than put every single teensy thing exactly back where it belongs the minute they're done with it. I've heard that the amazing Martha Stewart has created a line of paints based on the tints of the eggs laid by her Araucana hens. I wonder, would she be interested in a line of less muted hues based on the molds I found growing on the end of the loaf of bread this morning?

All right, I had straightened up a bit. Tossed out the bread, made the beds. I come from the South and am therefore genetically incapable—my husband says—of receiving company without making at least a little fuss. But when the reporter arrived (an hour late), I decided to take her out to the garden. It was already getting on toward suppertime, and besides, if you want to know why a zebra has stripes, you should look at it in the tall grass, right? My garden always looks great—I win nearly every valiant struggle there. The vegetables are forever trying to get out of hand, the melons want to squash, and the spinach tries to bolt, but

I keep them in line. I unlatched the gate and we walked into an early-autumn paradise shaded by my plum and Asian pear trees. Squash and gourd vines winked their yellow-eyed blossoms at us from the green wall of foliage that climbed the high fence. Long, elegant dipper gourds hung down from the trellis over our heads in a graduated array, like God's wife's measuring spoons.

While I harvested a basket of tomatoes and armloads of basil for pesto, I tried to explain that this, for example, was one of the things I am about. I'm someone who grows vegetables for her own table, not just to pass the time but as a kind of moral decision about how I want to live. She interrupted suddenly, asking, "Do you, like, meditate out here?" I backed up a few steps and explained that no, I, like, grew food out here. That as a matter of fact, we were going to be eating this stuff in a little while. She peered out over her sunglasses and asked, "Why? Don't you have stores out here?" And I backed up a few more steps, and then a few more, and pretty soon I just gave up. "Out here" was her code for incomprehensible territory, and that was where I lived. My kids came in and all of us had dinner together, and we treated our guest with all the deference and consideration one can offer a person who has shown up at the door hungry but doesn't speak or understand the language.

She went back to the big city and reported that I am not very open with strangers, have quaint ideas, and pay too much attention to my kids. I learned that Georgia O'Keeffe was right: The front and the back are all you should ever show to a person who doesn't really want the inside.

Most of the time I go right on growing tomatoes and basil and broccoli simply because they are good, we like them, I'm determined to figure out the right planting time for cole crops, and broccoli attracts hordes of green looper caterpillars that throw Lily's chickens into paroxysms of chicken joy. I do it because the world has announced to me, loudly, that it's time to make a choice between infinite material entitlement or a more modest, self-reliant security, and this is a step I can take in the right direction. Most of the time I raise up my wonderful daughters to have what I hope will be a useful blend of smart-aleck acuity and politeness, and once in a while we go down to help out the homeless shelter or dig a community vegetable garden because I want my kids to understand that compassion involves not just the heart but the hands. I write my poems, my congressmen, my letters to the editor, and I go on believing as I do, whether it makes any sense from the front and the back or not.

But like anyone else I am liable to be misunderstood, or scolded for standing apart from the crowd. I'm just one of a multitude of writers who venture outside the approved current of opinion *du jour* to get a better view of the complex struggle to reconcile cultural, national, and moral imperatives. Inevitably, some extremists will not tolerate this kind of art or dialogue. I've been called all the predictable names and some unpredictable ones; I've been misquoted in inflammatory ways by hate radio and its print equivalent in an attempt to impugn my patriotism and scare away readers. The historical mode of attack on writers (which continues into the present) is to avoid discussion of our actual ideas and instead declare us un-American for fabricated reasons and pronounce direly that no one had better listen to us, they'd best play it safe and just hate us. Inevitably, a few citizens will oblige: Some irate souls have vowed to uncover my true identity(!). Some are praying for my immortal soul, and two have offered to buy me a one-way ticket out of the country. (If I used them both, where would I end up?) I accept these gifts with the understanding that these people haven't the faintest idea who I am. It's important and worth noting here that the vitriolic mail almost never comes from anyone who has *read* me, but only from those who've read *about* me. It seems a certain sector has been led to associate my name with treason and sedition. Wow. The public may expect a circus, and fireworks—as Mark Twain wrote in bold-faced type on a handbill announcing one of his lectures—"in fact, the public may be invited to expect whatever they please." But they'll find no treason or sedition at my house, and they've rather pathetically missed my point, which is that it's *love* for my homeland that obliges me to participate in the discussion of preserving its integrity, and to take any risk necessary on my country's behalf. Otherwise, believe me, I'd live a safe and happy life writing cookbooks, or better yet, just cooking. It seems bizarre that a firm dedication to peace and the goodness of life should draw violent ire, but it does. Think of Gandhi, of Martin Luther King Jr. I'm hardly a drop in this river of tears and belief. Sometimes my heart catches in my throat and I just have to stop for a second with my hand on a doorknob or the cold metal of a key, assemble in my heart the grace of all we have to believe in, and say my own prayer for us all—that we will find the way through each hour of our lives that will have been worthy of the task.

In the long run I find it hardest to bear adversaries on the other end of the spectrum: those who couldn't care less, who won't or can't fathom the honest depths of love and grief, who opt out of the bull-ride through life in favor of the sleeping berth. These are the ones who say it's ridiculous to imagine that the world could be made better than it is. The more sophisticated approach, they suggest, is to accept that we are all on a jolly road trip down the maw of catastrophe, so shut up and drive.

I fight that; I fight it as if I were drowning. When I come down to this feeling that I am an army of one standing out on the broad plain waving my little flag of hope, I call up a friend or two and offer to make dinner for us. We remind ourselves that we aren't standing apart from the crowd, we *are* a crowd. We're a prairie fire, a church choir, a major note in the American chord, and the dominant one in the song of the world: a million North American students rejecting the tyranny of the logo and the sweatshop behind it; a thousand farmers in India lying down on their soil to prevent its being seeded with a crop that would steal their history and future; a hundred sheep farmers in southern France defying a fast-food hegemony by making cheese in limestone caves exactly as their great-grandparents did; tribal elders from east to west inviting peace to enter the world through its Hopi cloud dancers and its Sufi dancers; the Women in Black who stand in eloquent silence on every continent, refusing the wars that would eat their sons and daughters alive. We're the theater of the street, the accurate joy of children's hearts, the literature of tomorrow's wisdom arrived today, just in time. I'm with Emma Goldman: Our revolution will have dancing—and excellent food. In the long run, the choice of life over death is too good to resist.

When all else fails and I forget this, on those late nights when all the lights have gone out on my soul, I go into my office and read the *other* mail, the piles of love notes that outnumber the hateful letters two hundred to one. (Why does praise go in one ear and out the other, for so many of us, while we memorize criticisms verbatim? For the same reason the radio plays two hundred songs about loneliness for every one about family reunions. We hang our hats on heartache.) I am sustained by the kindness of strangers, who often send me remarkable gifts from the blue: a watercolor painting of a beloved bookshelf, a bar of handmade soap scented with rosemary, an exquisite book on the silk moths of North America, some precious tale of wonder or kindness, or just the perfection of gratitude, simply expressed. I can't possibly feel alone when so many—from prisoners to presidents, but mostly just everyday people—have accepted my words into their lives as they would the companionship of a friend, who say to me quietly in the park or the grocery when I'm least expecting it, "Thank you. Keep writing." And so I will, and when I need my own lifeline of words I read Walt Whitman, George Eliot, John Steinbeck, Arundhati Roy—people who have understood how to look life in the eye and love it back.

I fight against the drowning, knowing I can never go into the swamp of cynicism because if I do, I may never come out again. I'm not put together that way. I have children who are more precious to me than my life, and every molecule in me wants to promise them we'll get through this. We won't blow up the world before they get a crack at doing all the things grown-ups get to do in this howling hoot of a party: stand on your own two feet, get your heart broken, get over it, vote, drive a car, not drive a car, get dog-tired doing something that makes you proud, play the radio station *you* want, wear your heart on your sleeve, dance on the table, make a scene, be ridiculous, be amazing, be stronger than you knew, make a sacrifice that matters, find out what you're made of, cook a perfect meal, read a perfect book, kiss for an hour, fall in love for keeps, make love, make a baby, stand over your own naked child weeping for dread and wonder at the miracle.

If I got to make just one law, it would be that the men who make the decisions to drop bombs would first, every time, have to spend one whole day taking care of a baby. We were not made to do this killing thing, I swear. Back up. It's a big mistake.

The public is invited to think what it pleases, but to call me naive would be flat-out wrong. I have lived in a lot of different countries. So many, in fact, that now on some occasions when I'm asked here to vote yes or no, I want to color outside the lines. I turn over the referendum to look on the back for option 3: "RESOLVED to live with a little less so we can all share in the safety of having enough." In many countries, they give you that option. Our leaders tell us that these problems of ours are insoluble except by force, and that we must cede certain casualties to poverty and violence, and yet nearly every problem has already been solved by someone, somewhere: I've witnessed first-hand the blessedly kind health-care system of Spain, and I'd like to see ours follow its example. And the examples of Curitiba, Brazil, which recycles 70 percent of its trash, and Freiburg, Germany, which has brought back its streetcars and made automobiles unnecessary. Paris, Tokyo, and a hundred other municipalities have efficient public transportation systems that I'd like in my own city, thank you. I'd like an end to corporate welfare and multimillion-dollar CEO salaries so we could put that money into ending homelessness, as many other nations have done before us. I'd like us to consume energy, on average, at the modest level Europeans do, and then go them one better. I'd like a government that creatively subsidizes renewable energy and conservation, as Canada has done in some of its public school buildings, earning more than 100 percent return on the investment—which is returned again to the schools as equipment and teacher salaries. I would like us to ratify the Kyoto Protocol today and reduce our fossil-fuel emissions with the help of legislation that will ease us into safer, less wasteful, sensibly reorganized lives. I'd like to stake my pride on a nation that consistently inspires rather than bullies, that brings unconditional generosity to the table, and that dispenses justice over the inevitable bad deal with diplomacy and honor rather than with more bad deals. If this were the humane face we showed the world and the model we brought to working with it, every time, I believe our children might eventually be able to manage with a military budget the size of Iceland's.

There's a great big world out there, some of it clever and some of it frightfully cruel: I recently watched a film about women in Iran that made me go and kiss my sleeping daughters afterward, then kiss the ground on which we were lucky enough to be born. But there's no sense in getting cocky. England freed its slaves a century before we did, and right now many nations of the world consider U.S. policy to be awkwardly behind the times on many matters, from global conservation and science education to capital punishment. I find it helpful to remember this when I sometimes find myself outside the prevailing opinion here: My heart has independently found its way to a position that is, in a larger sense, *in*.

But whether I stand alone or with many, I'm still bound by that heart of mine to stand where we vote "none of the above" when presented with the equally odious choices of kill or be killed. I'm insulted by the suggestion that no other option exists, when nations all around us take very different strategies, many of them less belligerent than our own, with admirable success. I'm insulted further by the shallowness of the public debate, especially in wartime, founded as it seems to be on news reports devoid of any historical context. Our whole campaign against the Taliban, Afghan women's oppression, and Osama bin Laden was undertaken without nearly enough public mention of our government's previous involvement with this wretched triumvirate, in service of a profitable would-be pipeline from the gas fields of Turkmenistan. If the CIA and some U.S. corporate heads are romancing the same ilk elsewhere, right now, for similar reasons, then this high-minded talk of "Enduring Freedom" is wearing thin on my patience. The men in charge of our wars are well aware of these complex histories, but they speak to us in terms of simplistic threats without shades of cause or consequence, exactly as if we were all children.

We are expected to go along with this plan, in which people lose wars and corporations win them—the missile builders, the mining companies, the oil magnates, and that's just scratching the surface—and a little person like me should not dare be so insolent as to suggest a moment's time-out to review the monstrous human waste of an endless cycle of violent retaliation. Well, I'm daring. I have read that some of the missiles we are using (on the day of this writing) against our current enemy—one of the poorest countries on earth—cost a million dollars apiece. Excuse an outrageous suggestion, but has anyone considered just sending the innocent civilians the cash so *they* can dispatch the wretched tyranny in their midst and save everyone a huge cleanup? Masses of people tend to join cults of anger and vengeance only when they are desperate. History shows that populations with food in their bellies, literacy skills (women included), access to information, and immunization against the major diseases do not long tolerate martyrdom to the likes of the Taliban warlords or Saddam Hussein. And if those citizens were not grateful outright for our help in their liberation, at the very worst they might just forget about us—whereas our present strategy of asserting predominance with bombs is liberating some but starving others, driving millions to seek refuge in snow-covered, stony mountains, and ultimately sowing dragon's teeth of unforgettable enmity across the soil of one more desert.

Arrogance is a dubious weapon—an inappropriate side dish, anyway, to serve with a war. In fact, the very word *wartime* invokes for me a much more modest cultural mind-set, and lately I find myself saying this word quietly, again and again: *wartime*. It brings a taste to the root of my tongue, and to my inner ear the earnest tone of my parents recalling their teenage years. The word speaks of things I've never known: an era of sacrifice undertaken by rich and poor alike, of gardens planted and warm socks knitted in drab colors, communities working together to conquer fear by giving up comforts so everyone on earth might eventually have better days.

I went looking to see if I was imagining something that never happened. I found a speech delivered by Franklin D. Roosevelt on January 6, 1941, that made me wonder where we have mislaid our sense of global honor. "At no previous time has American security been as seriously threatened from without as it is today," he noted, as he could have done this day. But instead of invoking a fear of outsiders, he embraced their needs as America's own and called for defending, not just at home but on all the earth, what he called the four freedoms: freedom of speech and expression, freedom of religion, freedom from fear, freedom from want. "Translated into world terms," he said, the last meant "economic understandings which will secure to every nation a healthy peacetime life for its inhabitants." He warned that it was immature and untrue "to brag that America, single-handed and with one hand tied behind its back, can hold off the whole world," and that any such "dictator's peace" could not be capable of inspiring international generosity or returning the world to any true independence: "Such a peace would bring no security for us or for our neighbors. Those who would give up essential liberty to purchase a little temporary safety deserve neither liberty nor safety."

What reassurance I found in those words. I'm not an aberration, after all; I'm a good American, living in an aberrant moment, and I'm not the only one. When I ask around, almost everyone secretly agrees with me that we seem to be contriving a TV-set imitation—the look with no character inside—in our new-fab wartime of flags flapping above shopping malls and car-sales lots, these exhortations to purchase, to put down a foot and give not an inch. There's a rush on to squash the essential liberties of others and purchase some temporary safety. The four freedoms are not much in evidence. Faith and speech have taken hard blows, as countless U.S. citizens suffer daily intimidation because their appearance or their mode of belief or both place them outside the mainstream of an angry nation at war. Any spoken suggestion that there might be alternatives to violent retaliation is likely to be called an affront against our country. I have struggled to find some logical path that could lead to this conclusion—that is, the notion that ambivalence about war is un-American—and have identified as its only possible source a statement made by our president:

"Either you're with us, or you are with the terrorists." He was addressing nations of the world, but that "us" keeps getting narrower. If FDR's words were published anonymously today, especially those about force leading only to a "dictator's peace," FDR would get hate mail.

It's true. In our hour of crisis, no modern leader called on us for voluntary material sacrifice. The entitlement to personal gain is now, apparently, a higher value than duty to our country's greater good; please note that the wealthiest among us who rushed to dump their failing stocks and give our economy a black eye were never called unpatriotic. No leader could oppose it. No one dared us to put ourselves in the world's shoes (or its bare feet) and share, at least to some extent, in its fate. No public official even pointed out that we could improve our security immediately through our own collective action—by turning to local economies of production and distribution for our food and other necessities, by conserving energy, by turning off the TV and seeking solace from a city or national park or the hummingbird in the backyard instead of a new pair of shoes made in Malaysia. What could be better for our country, including its own economies, than to ease ourselves away from a framework of international profiteering that's proving perilous for so many reasons? But to call for this out loud might rattle the unassailable right to global moneymaking. It might be called treason, or sedition.

Such coldhearted values drive me back to my own faith as I mourn for the humane vision of a time that went before, and hope that vision will soon return to us. Freedom from fear, freedom from want—these clearly aren't meant just now for the Afghan civilians placed at risk of starvation. Our costly campaigns have put a notion of safety peculiar to ourselves ahead of any concern for the majority of world citizens who are starving and frightened—or for that matter, the hungry here at home.

Life takes awful, surprising turns; that's no news. I'm aware that just thirteen months after Roosevelt's eloquent call to conscience, the War Department persuaded him to order the internment of Japanese Americans. (The War Department, it's now known, manufactured threats of resident treachery to stir up public fear and uphold the concentration camps when they were challenged as unconstitutional.) But history's griefs can't entirely cancel its glories; there *was* that January day—the speech is archived as proof—when an American president proclaimed the lives of civilians on other soil to be as precious as our own. I would have planted a victory garden and accepted leaner rations to further that vision of a kinder world, in which all hungers mattered.

In fact, I'm planting one now: In response to September 11, a national network of gardeners has developed the means to devote a few rows of our gardens to the food banks that feed the hungry in our own communities. If our present leaders can't ask us for this sort of patriotism, we'll just go ahead without them. The public may expect fireworks.

After Roosevelt's famous speech, Norman Rockwell painted the four freedoms; his *Freedom from Fear* shows two parents in a darkened attic bedroom, tucking two little boys into bed. To look at that image now brings my thoughts to two other children, one nearby and one very distant. As our war drives a population into refugee status, immense waves of new recruits are entering schools in Pakistan and other places where young men train to a lifelong vow of vengeance against America. One, somewhere, is just a boy, the age of my younger child. Today that child and mine enter new lifetimes as hater and hated, and the door locks behind us all. The pacts begun today will long outlive the men in Washington and the momentary popularity of this war. Do those men really believe we have bombs enough to destroy every storefront or cement shell in the world that could serve as a school for hatred, when hearts are so turned? If my country's leaders can't tap into a vein of compassion right now, I ask them to search out prudence. I am the parent tonight in that darkened bedroom, with my knuckle to my mouth as I look at these children. I wish to reinstate Roosevelt's plea for a worldwide reduction of armaments, "in such a thorough fashion," as he said boldly—yes, in wartime—that *no nation* "will be in a position to commit an act of physical aggression against any neighbor, anywhere in the world."

My parents undertook wartime as a submission to sadness, not an indulgence in glory. They were led through it by a man who spoke with a heart full of intelligent remorse. I wonder what's happened to leaders who saw enduring peace as a house built on right, not on might, and knew that the world could never be right until all its people were free from hunger, censorship, and the dread of bombs. I wonder where they are now, all the teenagers and adults of a great generation who threw their hearts into an era of living simply, that others might simply live. I wonder if anyone else is feeling the hollow ring to this loud new wartime motto, "We'll show our enemies we're more powerful than they are." Our enemies know that already; they've known it all their lives, as they trained to the careful, hateful mastery of the tools that the weak may use against the mighty. They can plainly see we are richer, stronger, in every way more capable of destruction.

I would like us to show them, instead, that we are better.

The end of innocence for America, and the dawning of a better time, will both be ours when we've come to grips with an awful, irrevocable, wondrous truth: The biggest weapons we'll ever build cannot ever really make us safe. Believe it. When there are people on earth willing to give up their lives in hatred and use our domestic airplanes as bombs, it's clear that we can't outtechnologize them. We can't beat cancer by killing every cell in the body—or we *could,* presumably, but the point would be lost. We have been drawn into a protracted war of who can hate the most. There is no limit to that escalation. It can end only if we can summon courage enough to say it can't possibly matter who started it, and then begin to try and understand, and alter, the forces that generate hatred.

Horrifying enemies must be stopped, their violent plots intercepted, yes. But to write them off as unworthy of our study and comprehension is a pious and fatal mistake. Every step can be retraced. The terrorist network now known as Al Qaeda, with its horrific animosity toward the United States, has grown out of nearly fifty years of history. Many nations, and several multinational oil companies, have had roles in this morality play. The cast contains few heroes, many villains, and some genuine wild cards. After so much time, there is no possible system of accounting complex enough to determine genuine justice here; not everything can be forgiven or made right, and to suggest that the loss of innocent lives can ever be compensated is demeaning to life. Not all men of this world may be made to see eye to eye. But history shows they may often be induced, by mutual consent, not to put each other's out.

This crucial passage to understanding must begin somewhere. And religious tolerance may somehow be introduced into every discussion in which anyone currently claims God on his side. Every culture has its pride and prejudices, to be sure, but any that claim to own God are forgetting He started out as a most ordinary, everyday celebration of universal human dignity. If we are tempted, even subconsciously, to absorb the shock of anti-American terrorism as evidence that the Muslim world is irrational, violent, and undemocratic, we ought to remember that same world first invented and gave us the scientific method; hospitals; paper books as an instrument of widespread knowledge; our numbers; and the systems of algebra and trigonometry that make possible such feats as the design and construction of tall buildings. Terrorism against the United States is unnatural to the Islamic faith, as surely as the eleventh-century crusaders were failing to carry the true spirit of Christianity into Jerusalem when they boasted, "We rode through the blood of the Saracen up to the knees of our horses!" God must be very tired, by now, of being dragged into godless assaults on human flesh.

We should be tired of it, too. My country has been at war, secretly or openly, for virtually every year of my life, though my fellow citizens and I were mostly insulated from what that really felt like until September 11, 2001. Then we were chilled to our spines, and we began to say, "The world has changed. This is something new." We are right to weep aloud for this devastation; we should raise our forearms to the unsheltering sky and weep forever. But if there really is something new under the sun in the way of war, some alternative to the way people have always died when heavy objects are dropped on them from above, then please in the name of heaven I would like to see it, *now*.

On my desk sits a small black-and-white portrait of the world in a new year, when the year was 1903, that graced the cover of Emma Goldman's magazine *Mother Earth,* and words of hers that have crossed a century to reach me: "Out of the chaos the future emerges in harmony and beauty." Promises and prayers contain their own kinds of answers, as consecrated aspiration. I need this one now, as I need air and light.

I don't know what lies around the bend for us. I'm as scared as anybody, and grieving already: the end of nature and biodiversity, of safety and the privilege of travel; we have such larger losses to ache for than the end of the SUV as we know it. We may already be looking at the end of the world, in the form we least expect. It would be a pure, hellish irony of history if the same smallpox germ that was let loose on this continent two hundred years ago by the European arrivals, which quickly killed some 98 percent of the indigenous American population, were to revisit us again with the same results. It does not seem safe to assume we will ever know the moral of our story.

What I can say for certain is that many things will change for us, and fairly soon. We've built our empire on the presumption of endlessness for certain resources, which we are now running out of: more forests, more easily exploited oil, more economic growth based on more untapped markets for our goods. Alas, the nomads in Lorena Province may already be buying as much Coca-Cola as they're ever going to be induced to want. "The time will soon come," writes Wendell Berry, great prophet of our age, "when we will not be able to remember the horrors of September 11 without remembering also the unquestioning technological and economic optimism that ended on that day. This optimism rested on the proposition that we were living in a 'new world order' and a 'new economy' that would 'grow' on and on, bringing a prosperity of which every new increment would be 'unprecedented.'"

Every time I read an argument justifying further oil drilling in sensitive places, I notice that it begins with the caveat, "Unless Americans are willing to accept a drastic lifestyle change." As if that were the one thing that could never happen. As if many new kinds of shortage weren't already on the docket, scheduled for arrival, *period*, before my kids get to be my age. Scientists have been trying gently to remind us that the "fossil" in fossil fuel is not a metaphor or a simile. That oil is going to dry up eventually, and no political voodoo can induce dinosaurs or prehistoric fern forests to lie down and press themselves into more ooze for us on the timetable we require.

The writing has been on the wall for some years now, but we are a nation illiterate in the language of the wall. The writing just gets bigger. Something *will* eventually bring down the charming, infuriating naïveté of Americans that allows us our blithe consumption and cheerful ignorance of the secret uglinesses that bring us whatever we want. I am not saying I'm in favor of the fall; it terrifies me. I'm saying when the nine-hundred-pound bear gets all the way out to the *very* tip end of the limb, something's going to crash. Nostalgia for an earlier ignorance is not the domain of this discussion. Sitting here eating as fast as we can, while glancing around for the instrument of our demise, isn't it either. Would that the instrument might be a reconstruction guided by our own foresight and discipline, rather than someone else's hatred.

To wage war is human nature, I'm told, and the only way to settle a shortage of resources. I don't buy it. There is Jason's swashbuckling approach to the dragon's-teeth warriors, and there is Medea's more intuitive one, and both—for the record—are human. When the most recent round of bombings began, my mother and I declared to each other, "When the going gets tough, seems like men reach for a weapon and women look in the pantry." (My apologies, and deepest thanks, to you guys who were standing right there with Mom and me at the pantry door.) Slightly more than half of us down here on earth are of the pantry persuasion, and we didn't all of us get here by being efficient killers. By *here* I mean in charge of the place, numbering in the billions and wreaking our will on the planet. We got here by being social animals, communicative animals, cooperative animals, bipedal animals, tool users, seed savers, cagey mate choosers, bearers of live, big-brained young who seem determined each time around to outsmart their parents' generation, and frequently do. We are much too clever an animal, it seems to me, to kill ourselves now.

This is the lot I was cast, to sit here on this sharp, jagged point between two centuries when so much of everything hangs in the balance. I get to choose whether to hang it up or hang on, and I hang on because I was born to do it, like everyone else. I insist that I can do something right, if I try. I insist that you can, too, that in fact you already are, and there's a whole lot more where this came from.

That manner of thinking does not seem to be the fashion at this sharp, jagged little point in time, where the power is mighty and the fashion is coolness and gloom and one raised eyebrow. But still I suspect that the deepest of all human wishes, down there on the floor of the soul underneath the scattered rugs of lust and thirst and hunger, is the tongue-and-groove desire to be understood. And life is a slow trek along the path toward realizing how that wish will go unfulfilled. Such is the course of all wisdom: Others will see the front and the back, but inside is where we each live, in that home where only one heart will ever beat. There we have to make our peace with all we need of sorrow, and all we can ever know of the divine, by whatever name we can call it.

What I can find is this, and so it has to be: conquering my own despair by doing what little I can. Stealing thunder, tucking it in my pocket to save for the long drought. Dreaming in the color green, tasting the end of anger. Don't ask me for the evidence. The possibility of a kinder future, the existence of God—these are just two of many things that fall into the category I would label "impossible to prove, and proof is not the point." Faith has a life of its own.

Maybe the cynics are on top of the game, and maybe they're not. Maybe it doesn't cost anything to hope, and those of us who do will be able to live better, more honest lives as believers than we could as cynics. Maybe God really is just a guy on the bus. Maybe those really are his wife's measuring spoons hanging up there on my garden trellis, waiting to dole me out a pinch of grace on the day I need it. Maybe life doesn't get any better than this, or any worse, and what we get is just what we're willing to find: small wonders, where they grow.

Acknowledgments

The first person who believed this book should be assembled in some form, by me, was David Csontos. Frances Goldin and Terry Karten quickly agreed, and in turn convinced me, though I was at best the fourth person to get behind the project. (None of them but me, however, is responsible for any problems you may find with the result.) It took me a while to understand how lucky I was to have been handed a task to help keep me reasonably sane and focused through a frightening time. My friends and extended family also provided sanity and focus—especially Steven, my bedrock of solace and partner in both the practical logistics and the improbable dreams. I thank him, and thank my children, for being brave when we've all had to be, and inspiring me to take the necessary risks for what we believe in.

Fenton Johnson provided helpful comments on the manuscript, as did Sydelle Kramer, David Csontos, and Matt McGowan. Terry Karten was an author's best dream of an editor; Dorothy

Straight's copyediting was precise and inspired. Emma Hardesty showed loyalty and courage beyond the call of duty for an office manager, but if you knew her you'd expect nothing less. Frances Goldin, as always, guided me safely through the storms.

Most of my information on world poverty and other international humanitarian concerns came from various agencies of the United Nations. The quote from Barry Commoner, and information about the failure of the central dogma underlying genetic engineering, came from his excellent article "Unraveling the DNA Myth" in *Harper's,* February 2002. Natsuki Gehrt and David Csontos advised on matters Japanese, and for what I may still have gotten wrong, *sumimasen!* Larry Venable and colleagues at the University of Arizona provided information about seed banking for "Called Out." Information about cats and songbirds in "Setting Free the Crabs" came from R. Stallcup, "A Reversible Catastrophe," in the *Observer,* 1991. Other sources of information are listed in the foreword.

Paul Mirocha worked beside me, sometimes literally, to provide the illustrations and cover art for these essays as I was writing them. My book's mind gained its face, and a lovely new sense of itself, through his remarkably thoughtful collaboration.

To the organizations that will receive royalties from this book, I give my thanks for your supportive presence in my own life and your important work in the hopeful reconstruction of a better world. I urge every reader to maintain that gentle reconstruction in your own communities, as well as supporting these and many other national organizations doing similar work: Physicians for Social Responsibility (www.psr.org), Habitat for Humanity (www.habitat.org), Heifer International (www.heifer.org), and Environmental Defense (www.environmentaldefense.org).

I'm deeply indebted to the readers, booksellers, librarians, and friends who stood by me through the months when a handful of ultraconservatives sliced part of a sentence from my essay in defense of the flag, reversed its meaning, and paraded it across the country to revile me as Patriotically Incorrect. My readers, who understand patriotism to be a far nobler vocation than gossip-mongering, responded to the attacks by buying my books in great numbers. I will never forget this, or cease my effort to live up to your faith in me.

Finally, my mother never once told me not to stick my neck out. She gets the Maternal Medal of Honor.

Some of the essays in this book appeared previously in slightly different forms in the following publications: "Seeing Scarlet" in *Audubon* and *The Best American Science Writing 2001,* E. O. Wilson, ed., Houghton Mifflin; "The Patience of a Saint" in *National Geographic*; "Called Out" in *Natural History*; "Saying Grace" in *Audubon*; "Letter to My Mother" in *I've Always Meant to Tell You,* Constance Warloe, ed., Pocket Books, 1997; "Going to Japan" in *Journeys*, PEN/Faulkner Foundation, 1996; "Life Is Precious, or It's Not" in the *Los Angeles Times*; "Knowing Our Place" in *Off the Beaten Path,* Joseph Barbato and Lisa Weinerman Horak, eds., North Point Press, 1998; "What Good Is a Story?" in *Best American Short Stories 2000,* Barbara Kingsolver, ed., Houghton Mifflin; and "Taming the Beast with Two Backs" (under the title of "A Forbidden Territory Familiar to All") in the *New York Times.*

"Stealing Apples" originally appeared as the introduction to *Another America/Otra America,* poems by Barbara Kingsolver, Seal Press, 1998.

Brief portions of the essays "Flying," "And Our Flag Was Still There," "Lily's Chickens," "Marking a Passage," and "God's Wife's Measuring Spoons" first appeared as op-ed pieces in the *Los Angeles Times, San Francisco Chronicle, Washington Post, Boston Globe, Arizona Daily Star,* and *New York Times.*

The quote from Emma Goldman is from "The New Year," *Mother Earth* magazine, January 1912, used with permission, The Emma Goldman Papers Project, http://sunsite.Berkley.edu/Goldman.

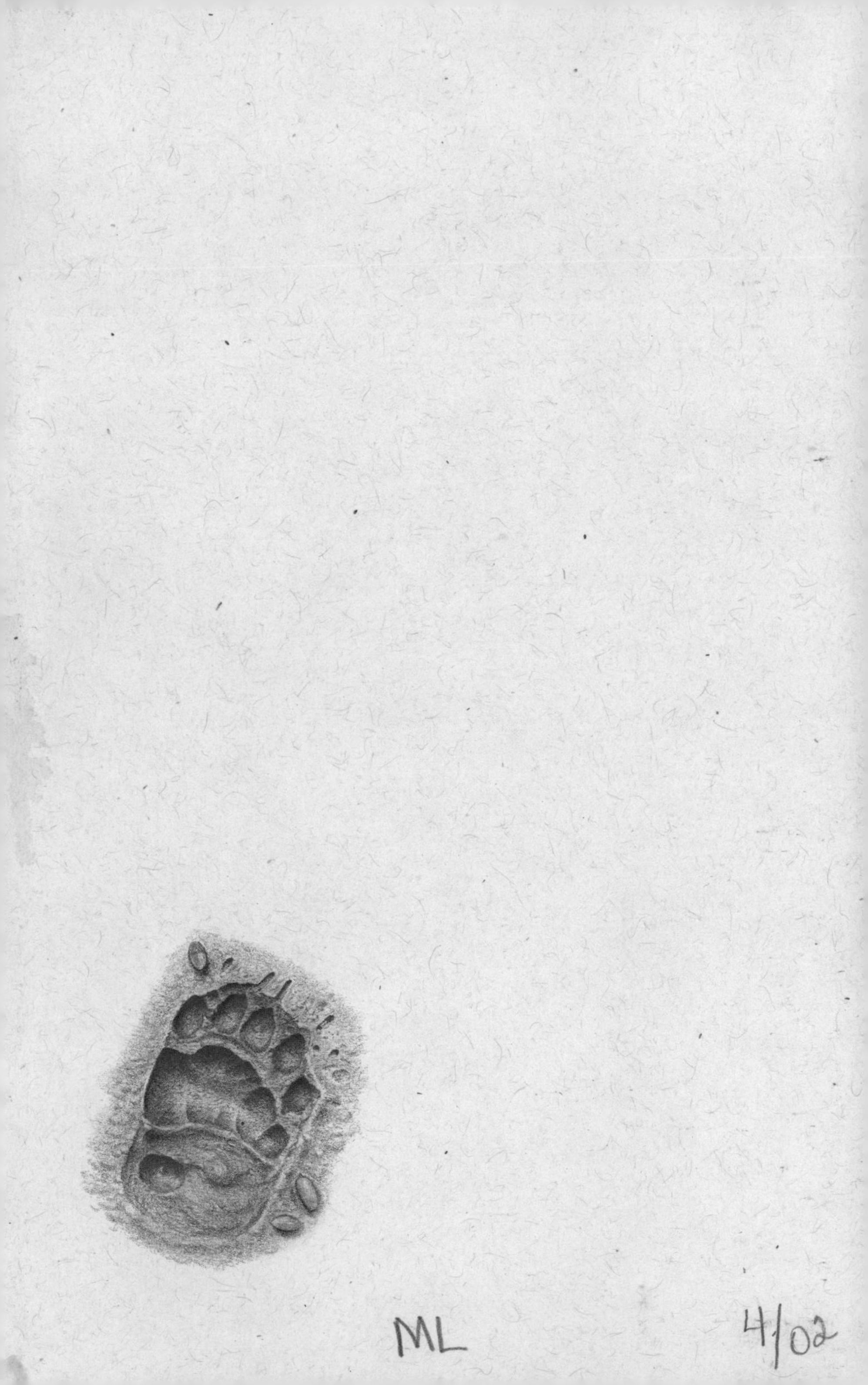

ML
4/02